WHY DO MEN
HAVE NIPPLES?

WHY DO MEN HAVE NIPPLES?

Hundreds of Questions You'd Only Ask
a Doctor After Your Third Martini

Mark Leyner and Billy Goldberg, M.D.

THREE RIVERS PRESS • NEW YORK

Published in the United States by Three Rivers Press, an imprint
of the Crown Publishing Group,
a division of Random House, Inc., New York.
www.crownpublishing.com

Three Rivers Press and the Tugboat design are registered trade-
marks of Random House, Inc.

Library of Congress Cataloging-in-Publication Data
Leyner, Mark.
    Why do men have nipples? : hundreds of questions you'd only
ask a doctor after your third martini / Mark Leyner and Billy
Goldberg.—1st ed.
        p. cm.
    1. Medicine—Miscellanea. I. Goldberg, Billy.
II. Title.
    R706.L49 2005
    610—dc22

                                            2005004882

ISBN 1-4000-8231-5
Printed in the United States of America

Design by Kay Schuckhart/Blond on Pond

21 20 19 18 17 16

First Edition

CONTENTS

CHAPTER 2.   BODY ODDITIES

CHAPTER 3.   ALL YOU (N)EVER WANTED TO KNOW ABOUT SEX

CHAPTER 4.   CAN I TREAT IT MYSELF?

CHAPTER 5.   DRUGS AND ALCOHOL

CHAPTER 6.   BATHROOM HUMOR

CHAPTER 8. OLD WIVES' TALES

CHAPTER 9. GETTING OLDER

ACKNOWLEDGMENTS

Billy:
My sincerest thanks to: my family for supporting me when I rambled on about this idea for years, my friends for asking me hilarious medical questions, and my beautiful and wonderful wife, Jessica, for appreciating my quirks and giving me the strength to finally get this done.

Mark:
I want to thank my wonderful and wise friend Billy Goldberg, who did virtually *all* of the work on this book and still happily let them put my name on the cover next to his.

Together we would like to thank Amanda Urban and Jud Laghi at ICM; our editor, Carrie Thornton; and the entire staff at One Jefe Productions.

PREFACE

Billy Goldberg: How did Mark Leyner and I come to know each other and pursue this heroic project, *Why Do Men Have Nipples?*

This is a long dark tale, a quixotic quest. A journey of two friends attempting to accomplish a nearly impossible task. We are an unlikely pair. I am a New York City emergency room doctor and Mark is a successful novelist and screenwriter. Not exactly the perfect literary match, but, our paths crossed and the rest is history. . . .

It began one frigid, blustery night in a busy New York City emergency room. I had been thinking about doing this book for many years. I had compiled questions and pondered answers but was never able to fully steel myself for such a perilous exploration and actually write any of them down. I had just been hired as a medical adviser on the ABC medical drama *Wonderland*. This short-lived show was a realistic drama based on the daily lives in a psychiatric emergency room and a prison psychiatric ward. The show added an ER character and I was hired to integrate the medical ER reality into the show. It was my job to bring the writers into our world of chaos. Most movie and TV writers knew nothing of real hospital medicine and were taken aback by the controlled disorder and gore of the emergency department and my world of science and human suffering.

I had been told by one of the producers that my visitor for that shift was going to be Mark Leyner. I consider myself to be well read, but I had never heard of this Dionysian postmodern superhero (Mark's description). A quick Google search revealed that he had published many novels, one of which was called *My Cousin, My Gastroenterologist*. I also found that he had written a television pilot for MTV entitled "Iggy Vile, M.D." I was confused and had no idea what to expect, but I was intrigued.

That night, I was at the bedside of a patient, assisting a resident in the placement of a nasogastric tube, when the nurse told me that there was someone to see me. I took off my gloves, pulled back the curtain, and there was Leyner. Nothing in my medical career could have prepared me for the character I was about to meet. He had the heavily muscled torso of a Bulgarian weight lifter and the weepy histrionic temperament of a teenage girl. He was babbling to no one in particular as he scarfed down fistfuls of Skittles from a paper bag. It soon became apparent to me that Leyner wasn't like the other TV writers I had met. He was a medical autodidact with an astonishingly bizarre and encyclopedic store of arcane medical knowledge. Within five minutes Leyner had regaled me with the precise pharmacokinetics of jimsonweed, a Fijian folk remedy for cannibal indigestion, the history of turf toe and crotch rot, and the inexplicable prevalence of supernumerary testicles in Wilkes Barre, Pennsylvania.

I knew it was going to be an interesting night.

As I made rounds with Leyner at my side, the first new patient to arrive was an "EDP." This is the term that we use for an emotionally disturbed patient. He was wildly agitated and a dozen burly New York City EMS personnel and cops were barely able to keep him restrained on a stretcher. Mark and I hurried over to see him wide-eyed and ranting psychotically. He was screaming in Spanish and English, "I am Superman, motherfucker. Get me Jimmy Olsen. I am faster than a speeding bullet, more powerful than a locomotive." I stepped up to the bed with the goal of getting an IV in and calming Superman down. He screamed again, "I am Superman, goddammit, your medications won't work on me." Leyner, who had been coolly observing the scene with clinical detachment, popped some Skittles in his mouth and made a stunningly unorthodox suggestion. "Give him kryptonite." I know that as you tell a story many times it begins to get embellished, but I remember that these words and these words alone calmed the patient enough so that we could get the drip going and get him under control.

The night continued to be a strange mix of the bizarre and touching, and I left my shift feeling that this strange little man would somehow have a profound effect on my life.

Mark Leyner: Even though I come from a long line of lawyers—and, in this society, lawyers and doctors are like warring sects in the Balkans—I'd always had a deep fascination with medical matters. Whereas most boys would subscribe to *Sports Illustrated* and *Boy's Life*, I waited eagerly for the mailman to deliver *Annals of Gastrointestinal Surgery* and *Journal of the American Society of Investigative Pathology*. Most kids begged their parents for trips to Disney World. I annually implored mine to take me to the Mütter Museum in Philadelphia, which houses this country's most glorious collection of medical oddities, including conjoined fetal quintuplets in formaldehyde and the preserved remains of the world's largest colon. I did seriously consider becoming a doctor, until I went to Brandeis, that is. There I saw firsthand the future doctors of America. Bunch of whining, ass-kissing, unscrupulous, morbidly neurotic premed students. My fascination didn't die though—in fact it became my secret inner life resulting in my first book of fiction being entitled *My Cousin, My Gastroenterologist*.

So I guess that my obsessive inclusion of graphic medical detail in all my subsequent books and a script I wrote for MTV entitled "Iggy Vile, M.D."—featuring an ale-swilling football hooligan punk of a surgeon—is what prompted Peter Berg to invite me to write for a television hospital drama he'd created called *Wonderland*. I'd just begun working on my first *Wonderland* script, when Peter called me one night raving about this guy who worked in the ER—this guy Billy Goldberg—and I took it all with a grain of salt, expecting him to be a grown-up version of the Brandeis premeds, but I agreed to meet him anyway. It turned out to be a glorious night. Billy was not the simultaneously insipid and officious physician I expected at all. The night was a revelation. What I saw that evening was amazing—a Chinese chef hit in his head with a meat cleaver, a Russian guy who came in with his ear in a bag of ice after his Rottweiler bit it off, and of course Superman. But it wasn't just the voyeurism. There was an immediate genuine connection between Billy and me, and there was something deeply compelling about the way he responded to the human needs of the people he treated in that chaotic and bizarre environment.

Billy: Several days later I showed up at the *Wonderland* production offices and I apparently had attained a newfound level of credibility. Leyner had told the tales of our evening and I am sure he had added a little bit of writer's embellishment. I was introduced to some of the other writers whom I had not met and sat briefly in their lavishly decorated and organized offices to answer the mundane medical questions from their respective scripts. My next stop was Leyner, and when I entered his office, it was like entering a tomb. The room was almost devoid of decoration and furniture, and it had a monastic feel. Leyner was lying on his stomach typing rapidly on the keyboard of his laptop. He looked up and without a greeting said, "Tell me everything you know about Kluver-Bucy syndrome!" We proceeded to discuss the clinical findings of this rare neurological disorder that causes individuals to put objects in their mouths and engage in inappropriate sexual behavior—obviously, a disease that was irresistibly attractive to the likes of Leyner.

Although critically acclaimed, *Wonderland* was canceled after three episodes. Leyner and I continued our strong friendship. We would talk about each other's work, and I particularly enjoyed reading his scripts and adding my meager suggestions. We decided a collaboration was in order and pitched several TV pilot ideas together. Working with Leyner brought me back to my long-standing idea for a book of cocktail party medicine questions. Leyner greeted my invitation to work on this book enthusiastically.

Mark: I considered it for a moment or two and remember thinking, "Hey, I could make a shitload of money and do almost no work!"

Billy: I felt that I had offered Mark the opportunity to be the doctor he always wanted to be. I thought he was ready to share the burden of this project, and what you are about to read is the result of all our, well, no actually, all my hard work.

Enjoy.

INTRODUCTION

When you're at a cocktail party, someone inevitably asks you what you do for a living. If you say that you are a doctor, the barrage begins.

Soon you're looking at someone's mole, consulting someone else on his brother-in-law's painful flatulence, racking your brain to explain the etiology of your hostess's episodic vertigo, and that's just the beginning. You would think that after twelve years of rigorous training and sleepless nights, doctors would have all the answers. But no! Not so. The sad fact is that one of the medical establishment's great shortcomings is its failure to teach what the general public really wants to know about medicine.

This book is an attempt to rectify this unfortunate situation. Inside these pages we will begin to answer some of the medical questions that real people ask. Pressing questions such as "Why does my pee smell when I eat asparagus?" "Is it true when they say 'beer before liquor, never sicker; liquor before beer, never fear?'" "Is sperm fattening?" "What causes an ice cream headache?"

This is not a self-help book or a medical manual. It's a glimpse at some of the strange things that people want to know from their doctors, but are too embarrassed to bring up in the ER or during an office visit. They only seem to find the courage to ask these questions after their third martini.

That's when the party begins.

DISCLAIMER

What you are about to read is mostly true, as far as we know. But this book in no way should substitute for a visit to your doctor. Remember, doctors are trained professionals. Also, do not attempt to answer these questions yourself unless you are a mother. Mother always knows best.

WHY DO MEN HAVE NIPPLES?

CHAPTER 1

It's 10 P.M., and my partner in writing and crime, Mark Leyner, and I are late as usual, but the party is in full swing. We brought a bottle of Don Julio tequila, which Leyner sampled voraciously in the cab, insisting that it needed to be screened for industrial toxins. We enter the elegantly appointed

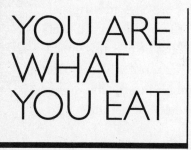

Park Avenue home of Eloise Cameron, a philanthropist, patron of the arts, and Botox junkie. Hors d'oeuvres are being served and the slightly inebriated and flush-faced Leyner grabs a mouthful of Swedish meatballs, kisses our hostess, and then comments, "Eloise, baby, better lay off the collagen. Kissing those lips is like making out with the Michelin man." She attempts to smirk with disdain, but the Botox leaves her face impassive.

I corral Leyner and we proceed into the living room. No sooner have we entered when I'm embraced from behind. I turn around and it's Jeremy Burns, an investment banker who sits two rows behind me at the Knicks games. Jeremy is well known to the Madison Square Garden food vendors for his insatiable appetite for hot dogs, cotton candy, and beer. He is now almost unrecognizable in his new Atkins-induced skeletonlike state. "Who exhumed you?" Leyner belches. I am overcome by embarrassment but secretly wetting myself with laughter. Jeremy tries to sidestep Leyner and as their arms brush, Leyner is covered with the grease that now oozes from Jeremy's pores. Leyner whispers to me, "This dude is all greased up like a rectal thermometer." I push Leyner away and he uses this opportunity to sneak over to the bar for another blast of Don Julio. I am left with Jeremy and his insufferable stories about life on the meat-and-fat diet, and a million medical questions about food.

If we are what we eat, why do we know so little about food and nutrition?

DOES IT REALLY TAKE SEVEN YEARS TO DIGEST CHEWING GUM?

What is it with seven years? You break a mirror, seven years of bad luck. Each dog year is seven human years. Seven years to digest swallowed gum? What if a dog broke a mirror then swallowed a pack of gum? Sounds like an algebra problem.

Chewing gum is not digestible but it definitely doesn't sit in your stomach for years. Gum actually might help things move through the bowels faster. Sorbitol is sometimes used as a sweetener in gum and this can act as a laxative. What does this mean? Yes, if you look carefully, you should see it floating next to all of those lovely yellow corn kernels.

WHY DOES YOUR PEE SMELL WHEN YOU EAT ASPARAGUS?

Asparagus contains a sulfur compound called mercaptan. It is also found in onions, garlic, rotten eggs, and in the secretions of skunks. The signature smell occurs when this substance is broken down in your digestive system. Not all people have the gene for the enzyme that breaks down mercaptan, so some of you can eat all the asparagus you want without stinking up the place. One study published in the *British Journal of Clinical Pharmacology* found that only 46 percent of British people tested produced the odor while 100 percent of French people tested did. Insert your favorite French joke here:_____.

....

3:32 P.M.

Gberg: Mr. Leyner, sir????

Leyner: Sir, reporting for duty, sir!

Gberg: You ready for a little work, son?

Leyner: Sir, permission to discharge my weapon into the sky, sir?

Gberg: Just don't hit the keyboard.

Leyner: What should we do today?

3:35 P.M.

Leyner: I have an idea . . .

Gberg: We have several things to do. Finish the preface, which we need to do together. Then we have 2 more intros.

Gberg: Or else we can add some funniness to some questions.

Gberg: You had an idea?

Leyner: Let's do that stuff (i.e., the preface and last two intros) . . . the real "writing" on Tuesday at your place . . . I think it works better with pacing.

Gberg: So let's work our way through the book.

Gberg: Let's start in the food chapter.

Gberg: We need to add something to this mother.

Leyner: Like?

Gberg: Some sidebars.

Leyner: I'm getting a Propel.

Gberg: Okay, so we need to add some expert medical commentary. By the way, do you think if we keep mentioning Propel, that delicious vitamin-enhanced beverage from the makers of Gatorade, we can get some free stuff?

Gberg: Only 20 calories per bottle. Sweet candy water!!! Where are you?

Leyner: OK, Pops. I'm here, eating my sandwich, drinking Propel . . . yes, absolutely!!!! We should shamelessly and unethically claim that Propel cures impotence, Crohn's disease . . .

Gberg: . . . and the smell of your urine from asparagus . . .

Leyner: . . . halitosis, and rectal whatever the hell you have.

Gberg: Should we add some French jokes?

Leyner: Certainly—let's claim that Propel deodorizes your urine EVEN after eating asparagus . . . then we'll get cases of the stuff!

Gberg: And you have to be a little less vulgar otherwise my wife won't be able to give this book to anyone as a gift without offending them.

Gberg: Bunch of puritans!

DOES SUGAR REALLY MAKE KIDS HYPERACTIVE?

Parents are always looking for an excuse to explain their children's bad behavior, and sugar has taken a lot of blame. This may come as no surprise, but the Coca-Cola Company doesn't want to take responsibility, and makes it very clear that studies have failed to find any substantial evidence proving a relationship between sugar consumption and hyperactivity. Well, the company is correct. Sugar does feed the body as an energy source, but it doesn't make kids hyperactive.

It is more likely that kids tend to eat sugary foods at times when they would be excited and rambunctious anyway (parties, holidays, movies, weddings, funerals). This can only be good news for the producers of such fine healthy treats as Cap'n Crunch with Crunchberries, Pixy Stix, cotton candy, and Laffy Taffy.

WHAT CAUSES AN ICE CREAM HEADACHE?

Aaaah, the joy of a Popsicle on a hot summer day.

One theory places the source for the brain freeze in the sinuses, where the pain may be caused by the rapid cooling of air in the frontal sinuses. This triggers local pain receptors.

Another theory postulates that the constriction of blood vessels in the roof and rear of the mouth causes pain receptors to overload and refer the pain to your head. There is a nerve center there, in the back of your mouth, called the sphenopalatine ganglion, and this is the most likely source of the dreaded ice cream headache.

A friend of ours suggested a quick cure of rapidly rubbing your tongue on the roof of your mouth to warm it up. Her demonstration included a bizarre clucking sound. Leyner tried this and found himself followed by a large goose of whom he seems to have become inordinately fond.

DOES EATING CHOCOLATE CAUSE ACNE?

For those of you who use chocolate as a substitute for sex, you can breathe a sigh of relief. There is no evidence that acne is caused by chocolate. Acne is connected more to changing hormones than to food choices.

Links have also been made between stress and acne. Recently, a group of dermatologists set out to prove that this common belief was also a myth but they found the reverse. Their study of twenty-two college students found that emotional stress was directly linked to acne severity.

But back to the chocolate issue, the University of Pennsylvania and the U.S. Naval Academy both demonstrated that chocolate does not cause acne. At the University of Pennsylvania, researchers fed subjects "chocolate" bars with no chocolate, while another group ate chocolate bars with nearly ten times as much chocolate as in a typical bar. Results of the experiment showed no significant difference in acne in either group. Other forbidden greasy foods like French fries, fried chicken, nachos, potato chips, and pork rinds probably don't cause dreaded zits either. So lighten up, kick back, and relax, and if that doesn't work go to McDonald's for a supersize fries and a chocolate shake.

WHY DO YOU CRY WHEN YOU CUT ONIONS?

Cutting an onion releases an enzyme called lachrymatory-factor synthase. This starts the process that leads to tears. This enzyme then reacts with amino acids of the onion and the amino acids are converted to sulfenic acids. The sulfenic acids spontaneously rearrange to form syn-propanethial-S-oxide, which is released into the air. When this chemical reaches the eyes, it triggers the tears by contacting nerve fibers on the cornea that activate the tear glands. Now you are crying.

Scientists have tried to make a "noncrying" onion but it seems that the crying enzymes are also responsible for the zesty onion flavor. But there may be some hope on the way. The group of Japanese plant biochemists that only recently discovered lachrymatory-factor synthase, the crying enzyme, believe that "it might be possible to develop a nonlachrymatory onion by suppressing the lachrymatory-factor-synthase gene while increasing the yield of thiosulphinate." Sounds delicious!

In the meantime there are several solutions to try to avoid the problem of onion-induced tears. Heating onions before chopping, cutting under a steady stream of water, or wearing goggles.

The most reliable: ordering takeout.

DO CUCUMBERS RELIEVE PUFFY EYES?

A well-placed cucumber may feel wonderful, but there is no special ingredient in it that reduces swelling under the eyes. Cucumbers are 90 percent water, and it is the cooling effect of the water that constricts the blood vessels around the eyes, therefore decreasing the swelling. The colder the cuke the better.

Some other swollen-eye solutions include black tea bags in cold water, the tannic acid content being the key to reducing swelling. Hemorrhoid cream also helps, but I'd prefer puffy eyes.

WHY ARE YOU SERVED JUICE AND COOKIES AFTER YOU DONATE BLOOD?

There is no solid medical reason for juice and cookies after blood donation. The idea is that this little snack will help to replenish your fluids and raise your blood sugar. But donating blood shouldn't really affect your blood sugar, and the small amount of juice that you drink probably has no significant effect on your fluid status. The best use of this snack is to allow you to rest and adjust before you go on your way after doing your civic duty.

Perhaps other food combinations could attract more blood donors:

1. For the upper-crust crowd: champagne and foie gras.
2. For the hipster: Vitamin Water and a PowerBar.
3. For the Atkins crowd: diet soda and a steak.
4. For the hip-hop gangster: a forty and some fried wings.

WHY DO WOMEN CRAVE CHOCOLATE DURING THEIR PERIODS?

There is little scientific support for a link between food cravings and the menstrual cycle. There have been suggestions
that chocolate cravings during menstruation are related to
a deficiency of magnesium or are linked to carbohydrate
consumption to self-medicate depression, but no strong
evidence has been found to prove either one. Studies have
placed volunteers on liquid diets that provided plenty of
calories and all the essential vitamins and minerals needed,
and participants still craved certain foods. This suggests that
nutritional deficits are not necessary for cravings of any kind
and that these desires are more psychologically based.

Medical texts, however, are filled with fascinating stories
about bizarre "food" cravings.

Pica is the medical term for a pattern of eating non-
nutritive substances (such as dirt, clay, paint chips, etc.) that
last for at least one month in the body. The name comes
from the Latin word for magpie, a bird known for its large
and indiscriminate appetite. Iron deficiency can cause pica
and can also cause a craving for ice, referred to as pago-
phagia. "Tomatophagia" has also been reported in a sixty-
six-year-old woman with iron deficiency who consumed
several whole tomatoes daily over a two-month period.
Her tomato cravings disappeared when her anemia was
treated.

WHY DO YOU GET BLOATED WHEN YOU EAT SALTY FOOD?

This is a common question that is most often asked by women who feel bloated because of PMS and believe that it is related to the amount of salt they eat. We both have learned over the years that you should never upset a woman if she is having premenstrual symptoms, so we went back to the medical school textbooks on this one to get the answer right.

Water accounts for 45 to 50 percent of the body weight in adult females and 55 to 60 percent of the body weight in adult males. Approximately 50 percent of this water is in muscle, 20 percent in the skin, 10 percent in the blood, and the remaining 20 percent in the other organs. Despite wide variations in dietary intake, the volume and composition of the body's fluids are maintained in an extremely narrow range as we lose (by urinating, sweating, etc.) as much water as we take in. In other words, the amount of a substance added to the body each day is equal to the amount eliminated or used by the body. This is called the balance state or the steady state.

Translation: if your kidneys are functioning normally, the amount of salt you eat shouldn't make you feel bloated. Maybe your pants are just too tight because you ate all that chocolate as a substitute for sex.

WHAT IS A FOOD COMA?

We are sitting at i Trulli, a top New York City Italian restaurant, and I have already unbuttoned my pants as I try to gather strength for dessert. I glance to my left and my sister-in-law has eaten herself to sleep. Her head is slumped on my wife's shoulder and drool is about to begin trickling from her mouth. After taking several pictures to add this event to family lore, I was again asked about the cause of the dreaded food coma.

There are many possibilities as to what causes the classic "food coma." Many people report drowsiness after eating the traditional Thanksgiving meal. Turkey is blamed for this soporific effect, specifically the amount of L-tryptophan contained in turkey. L-tryptophan is an essential amino acid and is a precursor of serotonin. Both serotonin and L–tryptophan have a calming, sedative effect in the human body.

L-tryptophan is naturally found in turkey protein but is actually present in many plants and animals, including chicken and cows. The average serving of turkey (about 100 grams or 3.5 ounces) contains a similar amount of L-tryptophan as found in an average serving of chicken and ground beef.

Two other factors that contribute to the desire to sleep at the dinner table are meal composition and increased blood flow to the gastrointestinal tract. Studies have shown that a solid-food meal resulted in faster fatigue onset than a liquid diet. The solid-food meal also causes a variety of substances to jump into action that ultimately leads to increased blood flow to the abdomen. This increase in blood flow and an increase in the metabolic rate for digestion can contribute to the "coma."

Now, I can tell the end of the family story. A good double espresso can sometimes be enough of a pick-me-up to get through dessert. But, in an attempt to resuscitate her comatose sister, my wife took her to the bathroom to splash water on her face and press her belly against the cold bathroom tiles. Unfortunately, time is the only true cure for the food coma.

WHY ARE YOU HUNGRY AN HOUR AFTER EATING CHINESE FOOD?

We fear that getting into any diet debate will cause us to be besieged by a gaggle of Atkins followers in a bacon-induced frenzy. But we may be safe this time, because the culprit may be carbohydrates—specifically, rice and pasta.

Chinese meals, for the most part, contain rice, little meat, and plenty of low-calorie vegetables. The rice and noodle dishes like fried rice and lo mein contain carbohydrates that cause the blood sugar to peak and then plummet, causing hunger. So, if you are going out for Chinese, don't forget the Peking duck, General Tso's chicken, or the spareribs. You may feel greasy and start quoting Mao, but you won't feel hungry later.

WHAT IS MSG, AND DOES IT CAUSE HEADACHES?

MSG is the sodium salt of the amino acid glutamic acid and a form of glutamate. Mmmm, doesn't that sound appetizing.

Glutamate is a naturally occurring amino acid that is found in nearly all foods, especially those high in protein. Monosodium glutamate (MSG) is used as a flavor enhancer in a variety of foods prepared at home, in restaurants, and by manufacturers of processed food. It is not fully understood how it adds flavor to other foods, but many scientists believe that MSG stimulates glutamate receptors in the tongue to augment flavors.

MSG has been the target of bad press based largely on reported reactions to Chinese food, the dreaded "Chinese Restaurant Syndrome."

For those who believe that they may react badly to MSG, the following symptoms have been reported:

- -

burning sensation in the back of the neck, forearms, and chest

numbness in the back of the neck, radiating to the arms and back

tingling, warmth, and weakness in the face, temples, upper back, neck, and arms

facial pressure or tightness

chest pain

headache

nausea

rapid heartbeat

**bronchospasm (difficulty breathing) in MSG-
    intolerant people with asthma**

drowsiness

weakness

- -

In 1958 the U.S. Food and Drug Administration (FDA) designated MSG as a Generally Recognized As Safe (GRAS) substance, along with many other common food ingredients, such as salt, vinegar, and baking powder, but consumers continue to have questions regarding MSG's safety and efficacy. However, there is general agreement in the scientific community, based on numerous biochemical, toxicological, and medical studies over the last twenty years, that MSG is safe for the general population.

CAN CARROTS HELP IMPROVE YOUR VISION?

The Roman emperor Caligula believed that carrots had the properties of an aphrodisiac, making men more potent and women more submissive. He is said to have fed the entire Roman Senate a banquet of only carrots so that he could watch the senators fornicate like wild beasts. This has nothing to do with eyesight, but it is quite a tale.

The carrot myth dates back to World War II when the British Royal Air Force was attempting to hide the fact that it had developed a sophisticated airborne radar system to shoot down German bombers. They bragged that the great accuracy of British fighter pilots at night was a result of them being fed enormous quantities of carrots. It is true that carrots are rich in beta-carotene, which is essential for sight. The body converts beta-carotene to vitamin A, and extreme vitamin A deficiency can cause blindness. However, only a small amount of beta-carotene is necessary for good vision. If you're not deficient in vitamin A, your vision won't improve no matter how many carrots you eat.

In fact, the ingestion of excess vitamin A can cause toxicity, which can include symptoms such as yellow-orange coloring of the skin, hair loss, weight loss, fatigue, and headache.

DOES COFFEE STUNT YOUR GROWTH?

I, Billy Goldberg, would like to dedicate this answer to my dear friend caffeine. He has been with me through good times and bad. Without him I would not have survived the long nights of my hospital residency nor the deadline of this book. To my friend I proclaim, "I do not hold you responsible that I am only five foot nine!"

Actually there has been considerable research on whether caffeine consumption is linked to osteoporosis. Overall, it can be concluded that moderate caffeine consumption is not an important risk factor for osteoporosis, particularly where women consume a healthy balanced diet. Some research suggests that regular caffeine consumption may lead to loss of calcium in the urine, but this does not have a measurable effect on bone density either. So as long as you have a balanced diet with adequate calcium intake, you can enjoy your espresso with no cause for concern.

So, why did our parents scare us with this myth when we wanted coffee as children? Probably for the same reason that they invoked the fear of losing an eye whenever we ran with scissors or snapped a towel. Pure parental mind control. It also helps if the child falls asleep and leaves Mommy and Daddy alone to find out if there really is a G–spot (see chapter 3, page 94).

WHY DOES SKIPPING YOUR MORNING COFFEE CAUSE A HEADACHE?

We truly are a nation of drug addicts. With alcohol, nicotine, and caffeine, we are constantly medicating ourselves to get through our daily activities. Now that people are commonly found freebasing caffeine in the form of Red Bull, we need an answer to this pressing question: Does cutting out the morning cup of joe cause a 4 P.M. headache from hell?

It is clear that caffeine can have an effect on headaches. Caffeine is present in both over-the-counter medications (Excedrin) and prescription medications. Caffeine acts to constrict blood vessels and therefore helps some headaches. But, the withdrawal symptoms you experience when cutting out your daily coffee are not as clear-cut as you may think.

A 1999 study in *The Journal of Pharmacology* challenged the assumption that stopping coffee causes headaches. When participants in this study were unaware of the caffeine-withdrawal focus, the frequency and severity of their symptoms were much lower and sometimes nonexistent. A recently released analysis concluded that there is a withdrawal syndrome when stopping coffee. Symptoms are thought to be worse if you consume more caffeine and then abruptly stop, although not everyone suffers the same withdrawal symptoms. Other symptoms include fatigue, drowsiness, irritability, depression, or trouble concentrating.

If you want to wean yourself off gradually, you can follow Mark Leyner's schedule:

Monday—double espresso

Tuesday—latté

Wednesday—single espresso

Thursday—Snapple iced tea

Friday—soy half-decaf mocha cappuccino

Saturday—a 12-ounce Coke

Sunday—beer (no caffeine and a wonderful breakfast treat)

WHY DOES SPICY FOOD MAKE YOUR NOSE RUN?

There is nothing quite like that rush you get when you mistake the wasabi for pistachio ice cream. But alas, this doesn't lead to nose running. That is because wasabi does not contain capsaicin, the extremely irritating chemical found in jalapeño or habanero peppers. Capsaicin is believed to stimulate central nervous system fibers that control the quantity and thickness of mucus and other fluids secreted in the nasal passages and stomach.

For you trivia nerds, heat in peppers is measured on something called the Scoville Scale:

- -

0–100 Scoville units includes most bell/sweet pepper varieties.

100–500 Scoville units includes pepperoncinis.

500–1000 Scoville units includes New Mexico peppers.

1,000–1,500 Scoville units includes Espanola peppers.

1,000–2,000 Scoville units includes ancho and pasilla peppers.

1,000–2,500 Scoville units includes Cascabel and cherry peppers.

2,500–5,000 Scoville units includes jalapeño and Mirasol peppers.

5,000–15,000 Scoville units includes serrano peppers.

15,000–30,000 Scoville units includes the Chile de Arbol peppers.

30,000–50,000 Scoville units includes cayenne and Tabasco peppers.

50,000–100,000 Scoville units includes chiltepin peppers.

100,000–350,000 Scoville units includes Scotch Bonnet and Thai peppers.

200,000 to 300,000 Scoville units includes ha-banero peppers.

Around 16,000,000 Scoville units is pure capsaicin.

The single hottest known pepper is the Red Savina habanero. If you think the jalapeño makes your nose run, the Red Savina will leave you wading knee-deep in a puddle of your own nasal secretions.

DOES SPICY FOOD CAUSE ULCERS?

No, spicy foods do not cause ulcers. Stomach ulcers can be aggravated by a nice dash of Tabasco sauce. Drinking alcohol, smoking, or experiencing stress can also make ulcers worse.

Most stomach ulcers are caused either by infection from a bacterium called *Helicobacter pylori (H. pylori)* or by overuse of anti-inflammatory pain medications such as aspirin or ibuprofen. The ulcers caused by bacteria can be treated with antibiotics and the others treated by an end to the pill popping.

DOES ARTIFICIAL SWEETENER CAUSE HEADACHES?

The artificial sweetener Equal and the food additive Nutra-Sweet are both aspartame. Approved by the FDA in 1981, this sweetener is hotly debated as the cause of everything from headaches to seizures. The debate rages on via the Internet and in the medical literature. The FDA and the Centers for Disease Control (CDC) both claim that this product is safe, but there are also many reports that show that headaches may be present as an adverse reaction in some patients.

There is no solid answer to the question of artificial sweetener causing headaches, but here are several things that are guaranteed to cause them:

1. Trying to help your child with math homework.
2. Telemarketers who call early on Sunday morning.
3. The map of red states and blue states.
4. Being stuck in traffic when the only clear radio station is playing an Ashlee Simpson marathon.

DOES LICORICE CAUSE HIGH BLOOD PRESSURE?

To begin with, it is important to understand that the delicious artificial strawberry or cherry product that we happily eat in movie theaters is not true licorice. True licorice is black and contains glycyrrhizic acid. Therefore we cannot answer the more important East Coast versus West Coast debate about whether Red Vines are better than Twizzlers.

Medical literature contains a great deal of information about the link between licorice and high blood pressure, and if you happened to be reading the English-language abstract of an article from the Norwegian journal *Tidsskrift for Den Norske Laegeforening* in 2002, you might have found out that "the active component of liqorice is glycyrrhizic acid, which inhibits the enzyme 11-beta-hydroxysteroid dehydrogenase. This enzyme promotes the conversion of cortisol to cortisone and is thereby responsible for the specificity of the mineralocorticoid receptor to aldosterone in the collecting tubules. Inhibition of the enzyme allows cortisol to act as the major endogenous mineralocorticoid producing a marked elevation in mineralocorticoid activity, resulting in hypertension, hypokalemia, and metabolic alkalosis." I can't understand why candy companies don't use this as a slogan. Imagine the catchy jingles, funny commercials, and booming sales of black jelly beans.

CHAPTER 2

I am

able to finally escape from the torture of Jeremy's food inquisition, and I look around and can't find Leyner anywhere. The bottle of Don Julio is missing and there is a trail of shrimp tails that leads to the elevator. I find him sitting in the hallway, playing Chutes and Ladders with

BODY ODDITIES

the neighbor's children, and devouring cocktail sauce with a straw. I try to get him back inside and he snarls, "Are you out of your mind? I'm down a hundred and fifty bucks." His bark is heard inside and several revelers come outside to watch the action. A crowd has formed around the game and Mark is becoming surly with the children as his losses mount. It doesn't help that the children are mocking him by singing "The Gambler" by Kenny Rogers. The tides turn and Leyner has soon wrestled the weekly allowances and the school lunch money from the kids, who disperse crestfallen while muttering to themselves. Triumphantly, Leyner rises and shouts, "Punk-ass suckers go crying to your mommy. We're going to bring this party back inside and play some strip Candyland." He pockets his winnings, swigs the Don Julio, and we are off.

Back inside, Wendy Thurston, a senior editor at Half-a-Dozen Ponds Press, has fallen victim to Leyner's shrewd, merciless gamesmanship. She is down to her bra, thong, and socks. As Leyner wins another point, she removes her left sock, revealing the most beautiful alabaster-hued foot and immaculately pedicured webbed toes. Teary-eyed, Leyner turns to me and in a choir boy's piping, soprano weeps, "I have found my Cinderella!"

This romantic outburst leaves the party in stunned silence, and then I'm again besieged by a slew of body-related questions. What is it about sideshow body oddities that awakens our most primal desires and curiosities?

IS IT BAD TO CRACK YOUR KNUCKLES?

As I, Billy, was sitting on the beach, relaxing and leafing through an old copy of the *Journal of Manipulative and Physiological Therapeutics*, I came across the answer to this age-old question. I also wish my father had known this, because maybe he would have yelled at my brother less. Cracking your knuckles is not as bad as people think. The usual argument is that knuckle popping causes arthritis. This does not happen. Chronic knuckle cracking may cause other types of damage, including stretching of the surrounding ligaments and a decrease in grip strength, but not arthritis.

So what causes the pop? The sound is produced in the joint when bubbles burst in the synovial fluid surrounding the joint. Really interesting, huh?

WHY DO SOME FOLKS HAVE AN "OUTIE" BELLY BUTTON AND SOME FOLKS HAVE AN "INNIE"?

I didn't have the answer to this question until I delivered my first baby. I always believed that you had an "innie" if the doctor tied a good knot, and if he didn't, you were cursed with that funny-looking "outie." Well, there is no knot tying at all. We just put on a clip, cut, and wait for the umbilical cord to dry up and fall off. It is all random.

Sometimes someone can develop an "outie" because they have a hernia at this site. This also has nothing to do with the doctor's Boy Scout skills. I have recently heard of plastic surgeons removing an "outie" for belly beauty. How sad.

One question that cannot be answered, however, is why some belly buttons collect so much lint.

WHAT CAUSES MORNING BREATH?

In Australia, the "poo fairy" comes at night to take a dump in your mouth. In England, they say a long night at the pub leaves your breath "tasting like the vulture's dinner." And a Scottish friend with a new Hawaiian bride reports that a late-night fridge-binge of haggis and poi will leave you with the worst morning breath of your life.

So, given all these tales, we should probably start with the anaerobic bacteria, the xerostomia (a fancy word for dry mouth), or the volatile sulfur compounds (which are actually waste products from the bacteria). All these combine to give you that wonderful get-up-in-the-morning feeling of garbage mouth.

Other things also contribute to this oral smorgasbord: medications, alcohol, sugar, smoking, caffeine, and dairy products.

But don't run off and have your tongue sandblasted; there are simple things that you can do to fight morning breath. Brush regularly (don't forget the tongue), floss, and drink plenty of water.

Gberg: I was just thinking that the more chaotic this is, the harder it is for Carrie to edit. It might even induce a seizure.

Leyner: That's funny!! I think tormenting her is always a good sort of compass for us when we're lost and floundering.

Leyner: What is a seizure, actually?

Gberg: Is that the way it usually works in the creative process? Is your genius always fueled by torment?

Gberg: Abnormal electrical activity in the brain, why?

Leyner: My creative process is fueled by a sense of Nietzschean aristocracy and a simultaneous feeling that I'm an abject fraud.

Gberg: I think everyone feels like a fraud. What about me, trying to answer these unanswerable questions?

Leyner: Coupled with torment and an overwhelming need to be loved and liked (even) AND horniness AND creditors calling ALL THE FUCKING DAY LONG. Didn't you go to medical school in Ingushetia? You are a fraud.

Gberg: Where the hell is Ingushetia?

Leyner: Directly east of Chechnya. Check MapQuest.

Gberg: Enough of your Chechen obsession. Let's talk about the book.

Leyner: I told you . . . with the amount of money we're getting paid for this book, Mercedes and I are getting a time-share summer dacha in Chechnya.

••

WHY ARE YAWNS CONTAGIOUS?

Here are several things we can be thankful are *not* contagious:

- -

drooling

nosebleeds

itching

seizures

farting

- -

That said, there are several theories for what causes yawns and why they are contagious. It was originally thought that people yawned to get more oxygen, but this appears not to be true.

The most common theory is behavioral. In an article examining contagious yawns, Dr. Steven M. Platek and others state, "Contagious yawning may be associated with empathic aspects of mental state attribution and are negatively affected by increases in schizotypal personality traits much like other self-processing related tasks."

Huh? I find myself yawning right now.

What they mean is that people are unconsciously imitating others when they yawn. Humans are not the only species that yawn. Yawning is seen in many animals, including cats, fish, and birds, although we don't know what a yawning fish looks like either.

WHY DO MEN HAVE NIPPLES?

Since our editor thought this question made the best title for this book, we racked our brains to come up with a hilarious, witty, and informative answer to this question. Our attempts proved futile, so, in order to finish this book so another brilliant title wouldn't go to waste, we went for the boring, straight scientific response. Sorry.

We are mammals and blessed with body hair, three middle ear bones, and the ability to nourish our young with milk that females produce in modified sweat glands called mammary glands. Although females have the mammary glands, we all start out in a similar way in the embryo. During development, the embryo follows a female template until about six weeks, when the male sex chromosome kicks in for a male embryo. The embryo then begins to develop all of its male characteristics. Men are thus left with nipples and also with some breast tissue. Men can even get breast cancer and there are some medical conditions that can cause male breasts to enlarge. Abnormal enlargement of the breasts in a male is known as gynecomastia. Gynecomastia can be caused by using anabolic steroids. So, if Barry

Bonds ends up coming to the old-timers game with a pair of sagging 44DD man boobs, then I think we will finally have our answer to the steroid controversy.

CAN YOU LOSE A CONTACT LENS IN THE BACK OF YOUR HEAD?

It is common for people to come into an emergency room because they can't find their contact lens. Sometimes it is found folded and tucked beneath the eyelid, but other times it is nowhere to be found. So where is it???

Probably on the bathroom floor at home. A little anatomy lesson: there is nowhere else for it to go.

Other commonly "misplaced" items that lead people to the ER: tampons, condoms, and car keys.

CAN YOU LOSE A TAMPON INSIDE YOUR BODY IF THE STRING COMES OFF?

This is a surprisingly frequent question, and often a reason women find themselves in the emergency room. Patients often come in either because they cannot remove the tampon or because it has disappeared and they don't seem to know where it went.

Time for another anatomy lesson. The vagina is a potential space, not a hole or cavity inside the body. The walls of the vagina are normally in contact with each other unless something is inserted between them. When something enters the vagina, the body makes room for it. At the end of this potential space is the cervix. Therefore, there is no place for the tampon to go. It cannot be lost inside that small area and you should be able to remove it, or it can be easily removed by any doctor. Often we find nothing inside, and that means you probably forgot you removed it. Leaving a tampon inside too long can put you at risk for a serious infection, so don't be embarrassed to ask for help.

IS IT TRUE THAT THE TONGUE IS THE STRONGEST MUSCLE IN THE BODY RELATIVE TO ITS SIZE?

Now, we are sure there are many possibilities as to why someone would need the answer to this question. We never asked our friend who asked this question why this was important, but surely she had her reasons.

Some sources do agree that the tongue is the strongest muscle per size, but the tongue is actually made up of four muscles. The heart has also been mentioned, but since it moves involuntarily and is mainly an endurance muscle, it doesn't really get to the heart of this question (bad pun intended).

The sartorius, which slants across the thigh to the knee, is the longest muscle in the body. As for the strongest, there are two other candidates, the masseter, used for chewing, and the gluteus maximus. By gluteus! Who knew that our asses were so strong!

Another tidbit for you trivia geeks, here are Billy and Leyner's two favorite ass-vocabulary words:

callipygian: having beautifully proportioned
buttocks

steatopygic: an extreme accumulation of fat
on the buttocks

WHY DO YOUR TEETH CHATTER WHEN YOU ARE COLD?

The body usually maintains a constant temperature of 98.6 degrees Fahrenheit. At this temperature the cells of the body work best. If there is any significant change in temperature, it is sensed by an area of the brain called the hypothalamus. When the body gets too cold, this center alerts the rest of the body to begin warming up. Shivering, the rapid movement of the muscles to generate heat, then begins. Teeth chattering represents localized shivering.

WHY DO YOU HAVE AN APPENDIX IF YOU CAN LIVE WITHOUT IT?

The appendix is a small pouch off the large intestine. The wall of the appendix contains lymphatic tissue that is part of the immune system for making antibodies.

Removing the appendix doesn't cause any harm because there are several other areas in the body that contain similar tissue—the spleen, lymph nodes, and tonsils. The spleen and the tonsils can also be removed.

Gberg: You were going to give me a little something to add to the appendix question, Why do you have one if you can live without it?

Gberg: Some expert Leynerisms on vestigial organs.

Leyner: God put certain internal organs in the human body for purely aesthetic reasons. They just look nice when the forensic pathologist opens you up.

4:05 P.M.

Leyner: How do we know yet what all the vestigial organs are? A lot of the organs that seem crucial now may seem vestigial pretty soon.

Gberg: Please explain to me what is so beautiful about the appendix. It looks like a little wet caterpillar.

Leyner: It's so subjective, though . . . a little wet caterpillar is beautiful . . . vulnerable, bespeaking the evanescence of life and the unbearable limpness of it all. I'm sure at some time, somewhere, the appendix had its moment, its evolutionary "15 minutes" of utility.

Gberg: What the hell are you bespeaking of?

Leyner: There was probably some predator that only ate people without an appendix so that gene flourished for a while. . . .

Leyner: Speaking of vestigial.

Leyner: What other supposedly vestigial organs are there?

Gberg: The tail.

4:10 P.M.

Leyner: Wouldn't body hair also be considered vestigial now, since we don't live naked out on the primeval savannah?

Leyner: I try to live naked on an inner sort of primeval savannah, but you know what I mean. Body hair is surely some sort of atavistic throwback too.

Leyner: What's the purpose of pubic hair or back hair or even hair on the head?

Gberg: Hold on, I have to look up "atavistic." I need a freakin' thesaurus.

Leyner: It's all economics. There's not enough of a service industry devoted to the appendix, so it's anathematized as "vestigial" . . . hair's cool, what with the waxing industry and salons and shampoos and conditioners, etc. etc.

Gberg: You know, I love the word "merkin," but do people really wear those?

Gberg: Why would anyone really want a pubic toupee?

4:15 P.M.

Leyner: The whole idea of a merkin is so great! I think this whole culture of plucked women is pretty perverse, actually. It's sexually infantilizing. Who wants a woman who looks like a six-year-old down there?

Leyner: How big can an appendix get?

Gberg: I know, but by the same token, you don't want someone who looks like a Yeti.

Gberg: Is that how you spell it?

Leyner: A Yentl?

Leyner: Yentl?

Gberg: No, the abominable snowman.

Gberg: Are you saying that Babs wears a merkin?

Leyner: Well . . . I know we're digressing a bit . . . but rather a Yeti than a glabrous, waxen thing.

Leyner: Streisand is famous for the merkin. Made by the finest Venetian merkin-makers.

Gberg: Let's not digress. We can get back to the book.

Leyner: From Yak scrotal hair.

Leyner: OK . . . back to the book.

....

ARE CANKER SORES CONTAGIOUS?

One of the great secrets of medicine and one of the things that doctors aren't quick to admit is that we often don't have all the answers. Canker sores are one of those cases. Canker sores, medically known as recurrent apthous ulcers, are the most common oral disease and something that many of us have experienced. They differ from cold sores in several ways. Canker sores occur inside the mouth while cold sores show up on the lips. Cold sores are caused by the herpes virus and are definitely contagious. The etiology of canker sores is still unknown, although scientists have spent a great deal of time searching for the answer. Studies have suggested that this inflammatory disease is a result of abnormal immune response directed toward the oral membranes. Several bacteria and viruses have also been investigated as the culprit, but none has been found to be responsible.

WHAT ARE GOOSE BUMPS?

It's all about the arrectores pilorum.

What, you say, are arrectores pilorum?

These tiny little hair erector muscles that contract and raise the hair follicles above the skin. These are goose bumps or goose flesh or chicken skin.

What causes them?

They start with a stimulus such as fear, cold, or the sight of yourself in the mirror after a night of vodka-induced debauchery. This causes the sympathetic nervous system to become activated. The sympathetic nervous system is responsible for the body's "fight or flight" response. This sends a message to the skin and activates those little muscles.

WHAT REALLY IS HAPPENING WHEN MY FOOT FALLS ASLEEP?

Saturday night palsy is a condition often seen in the emergency room, not a sequel to a John Travolta film. It is caused by the same mechanism that makes your foot fall asleep but is a tad more severe. Saturday night palsy is caused when someone, who is usually really wasted, is unable to move an arm or leg in response to the pins and needles caused when a limb "falls asleep." It can lead to temporary or even permanent nerve damage.

Here is what happens in normal conditions. When pressure is exerted on part of your leg or arm, several things occur. Arteries can become compressed, making them unable to provide the tissues and nerves with the oxygen and glucose they need to function properly. Nerve pathways can also become blocked, preventing normal transmission of electrochemical impulses to the brain. Some of the nerves stop firing while others fire hyperactively. These signals are sent to the brain, where they are interpreted as burning, prickling, or tingling feelings. It is these sensations, paresthesias, that alert you to move your foot. Shaking your foot releases the pressure and nutrient-rich blood flows back into the area and nerve cells start firing more regularly. The "pins and needles" feeling can intensify until the nerve cells recover. That is why it is painful when you try to "wake up" your sleeping limb.

Persistent numbness or tingling can be a sign of certain medical conditions, and in those cases you should see your doctor.

WHY DO YOU GET BAGS UNDER YOUR EYES WHEN YOU ARE TIRED?

Feeling exhausted? Wondering why you have bags under your eyes that make you look like Droopy Dog or John Kerry?

Lack of proper restful sleep seems to cause dark rings for reasons not properly understood. The skin around the eye is the thinnest found anywhere on the body, and this thin skin allows dark, venous blood to show through.

Dark rings around the eyes are a common problem. They appear to be genetic and can get worse as you age and your skin gets thinner. Adequate rest, good nutrition, and overall good health tend to make the circles less noticeable. You can also wear sunglasses all the time.

WHY DO YOU LAUGH WHEN TICKLED?

You definitely don't spend a great deal of time learning about laughter in medical school. I know that doesn't surprise you since physicians are such serious people. The closest they come to humor is the physiological study of laughter—gelotology. There is even a form of seizures called gelastic seizures that causes sufferers to laugh incessantly.

Laughter is a complex process that requires the coordination of many muscles throughout the body. Laughter also causes an increase in blood pressure and heart rate, breathing changes, reduced levels of certain neurochemicals, and a potential boost to the immune system. So, overall, it is very good for you.

Researchers have attempted to decipher the purpose of laughter and many believe that the reason for laughter is related to making and strengthening human connections, a kind of social signal. Studies have shown that people are thirty times more likely to laugh in social settings than when they are alone. Reports also suggest that the origins of laughter may predate human evolution.

So, what about the connection between tickling and laughing?

Well, this tickling-induced laughter is actually a reflex. Scientists don't fully understand how this works, but because you cannot tickle yourself, the reflex seems to require an element of surprise.

WHY DOES SWEAT STINK AND STAIN?

Have you ever used the expression "sweat like a pig"? Think again. Pigs don't sweat. Pigs don't have sweat glands, which explains why they have to wallow in puddles and mud to cool off.

As for us humans, we routinely sweat as a way of eliminating excess heat and maintaining a normal body temperature. The average person has 2.6 million sweat glands distributed over the entire body except for the lips, nipples, and external genitalia. There are two different types of sweat glands, eccrine and apocrine. These glands are different in size and produce different kinds of sweat. Eccrine glands are located all over the body. Apocrine glands are different because they are found mostly in the armpits and groin. They are larger and open into hair follicles. Though sweat is mostly water, it is the small amount of protein and fatty acids in the apocrine sweat glands that gives armpit sweat that wonderful milky or yellow color. It is also what causes it to stain.

Sweat itself is odorless whether it comes from the armpits or other areas of the body. The funk begins when sweat mixes with bacteria that occur naturally on the surface of the skin. This distinctive odor is called bromhidrosis— foul-smelling sweat.

• •

Gberg: I was going to add a New York cab driver joke to the "Why does sweat stink?" question.

Gberg: Any thoughts?

Leyner: What's the joke?

Leyner: I love jokes.

Leyner: What's the stinky cabbie joke?

Gberg: I don't know one, but the scents of a taxi are so rude.

Gberg: It's either that overwhelming air freshener or wretched body odor.

Leyner: See!! It's all economics . . . cabbies won't run the AC . . . so of course they're gonna stink—especially the ones who wear the Irish fishermen's sweaters and the Latex underwear in the middle of the summer.

Leyner: Air freshener is, to me, worse than the smell it's supposed to obscure . . . it just makes me think of what the person is trying to camouflage, so my mind creates an even greater fetid fiction.

Gberg: I don't know. It depends what scent you are talking about. The hospital has some particularly vicious scents that need covering, like . . .

Gberg: Butt pus and

Gberg: bloody stool, which . . .

Gberg: They both sound like punk bands.

Leyner: Isn't the smell of sweat supposed to produce certain subconscious (or conscious perhaps) sexual responses? And . . .

Leyner: What the hell is "butt pus"?

Gberg: Like a perirectal abscess or a pilonidal cyst—you drain them and the scent is horrible.

Leyner: Oh . . . that's not so bad.

Leyner: I've smelled that.

Leyner: I have a pilonidal cyst—a dormant one though.

Gberg: You always were scent obsessed.

Leyner: I met a girl at Brandeis who also had one, and we soaked ours together. That's true.

Gberg: Sitz baths.

Leyner: Fond memories of her.

Gberg: A sitz schvitz.

Leyner: Yes . . . sitz baths—we were young and idealistic.

Leyner: Isn't a pilonidal cyst somehow related to a vestigial tail?

Gberg: I don't know.

Leyner: That's a fucking simple medical question, and your answer is "I don't know"!!!!!!!!!!

Gberg: It brings back the original point of this book. They never teach you the obscure stuff that people actually ask.

4:30 P.M.

Leyner: My grandfather used to go to Hot Springs, Arkansas, for "baths." Or so he told my grandmother.

Gberg: I can describe in detail the technique for draining a pilonidal cyst or talk about marsupialization, when you sew down the sides.

Gberg: Nobody wants to know that.

Gberg: And then I get mocked by some pumped-up little writer who couldn't marsupialize his way out of a peper bag.

Gberg: Not a pepper bag but a paper bag.

Leyner: Do people ask you strange questions in the ER? Or are they too freaked out by having meat cleavers embedded in their heads to make small talk with you?

Leyner: You gotta explain that, dude!!

Gberg: Explain what?

Leyner: What's marsupialization?

Gberg: You cut open the cyst and sew down both sides so it doesn't come back. You create a little pouch.

Leyner: Maybe I'll get that! I'll have the ass of a kangaroo!

....

WHAT IS SNOT?

Phlegm, snot, spit, boogers, sputum—all different varieties of the same thing. These terms are used to describe different forms of mucus, a slimy material that lines various membranes in the body (called, of course, mucus membranes). Mucus is composed chiefly of mucins (lubricating proteins) and inorganic salts suspended in water. Mucus aids in the protection of the lungs by trapping foreign particles that enter the nose during normal breathing. Mucus also makes swallowing easier and prevents stomach acid from harming your stomach wall.

As for the different varieties, phlegm is one type of mucus. By definition phlegm is limited to the mucus produced by the respiratory system, excluding that from the nasal passages (that is what we refer to as snot), and that which is expelled by coughing (sputum). In medieval medicine, phlegm was counted as one of the four bodily humors, possessing cold and wet properties. Phlegm was thought responsible for apathetic and sluggish behavior, which is how we get the word *phlegmatic*. Boogers are less historical, a slang word for dried nasal mucus or snot.

The presence of mucus in the nose and throat is normal. When you are sick the mucus can become thicker and change colors. Color is not a clear indication of a bacterial infection, but persistent rust-colored or green mucus tends to indicate a more serious condition.

For those do-it-yourself types, there are many ways to make home mucus to prepare yourself for a career in medicine:

RECIPES

Ingredients

½–1 pound fresh okra

1–2 cups water (the less water you add, the thicker your mucus will be)

Instructions

1. Chop the okra into large pieces and place them in a saucepan with a tight-fitting lid.
2. Add water to cover and boil the okra, about 10 to 15 minutes, until it is a dark grayish green and very soft.
3. Turn off the stove and remove the lid. Let your slimy substance cool.
4. Strain the slimy mess into a bowl and discard the okra.

Or

1. Stir ⅛ cup borax into 500 ml (2 cups) warm water. It's okay if some borax remains undissolved. Allow solution to cool to room temperature.
2. In a separate container, stir 2 spoonfuls of glue (Elmer's) into 3 spoonfuls of water.
3. Stir a couple drops of food coloring into the glue mixture.
4. Add a spoonful of the borax solution to the glue mixture. Stir (if in a bowl) or squish (if in a Baggie).

WHAT ARE EYE BOOGERS?

To answer this question we called one of my smartest friends, an Ivy League–educated ophthalmologist who is a retina surgeon at a prestigious university hospital. He's the kind of guy who sends me Proust as a birthday gift. Doesn't watch TV. Listens to NPR. So, we go to him for the answer. . . .

Nothing. He tells me he will look it up. This just goes to show you that medical school sometimes misses the really simple stuff.

So, who has the answer? Honorary physician and expert on medical oddities Mark Leyner wrote about this malady in *Maximum Golf* magazine. Here, one pseudoschizophrenic golfer hears two golf announcers having the following discussion in his head:

• •

Announcer B: Michael's a bit off center—I'd say less than a foot from the left edge of the mattress and maybe a good foot and a half from the right rim. He's got his left arm tucked under the pillow—

Announcer A: Which looks to me like a 245-thread-count cotton-twill shell filled with a 95-percent-Canadian-feather-and-5-percent-down blend.

Announcer B: What's that in the corner of his left eye? A small emerald green particle. Can you make that out?

Announcer A: That's the mucopolysaccharide secretion from the lachrymal gland that's accumulated and crystalized overnight, Bobby.

Announcer B: Eye gunk. My mama used to call that a "sleeper."

Announcer A: Well, we've got a lovely aerial view of Michael's sleeper from the MetLife blimp, *Snoopy Two,* cruising at thirty-five miles per hour at an altitude of twelve hundred feet. Our thanks to Captain William Schmickling and his crew for that shot. Absolutely splendid.

Announcer B: Chris, he's gotta get that outta there. What would you do in this situation?

Announcer A: There's the very slightest breeze coming through the open window, but not sufficiently gusty to warrant any sort of major tactical adjustment. I'd use an index finger here, position it on the corner of the eye, precisely there at the lachrymal duct, and just ever so gently, ever so deftly, roll the particle out.

Announcer B: You can't try to do too much here.

Announcer A: Just get it out, actually—that's a job well done.

Announcer B: Reminds me of when Ernie Els got an eyeful of sandpiper guano at the AT&T Pebble Beach National Pro-Am in '95. Played the back nine basically half-blind. One of the most courageous exhibitions I've ever witnessed.

• •

This eye gunk is nothing serious. While you sleep, a mixture of oil, sweat, and tears collects near the corners of your eyes. As the tears dry up you get left with a nice little bit of crust.

WHAT ARE THOSE LITTLE HALF MOONS IN YOUR NAILS?

The pale half-moon shape at the base of each nail is called the lunule. It shows where the hardening process is not yet complete.

The American Academy of Dermatology provided these nail facts:

- **Nails grow about 0.1 mm (or about .004 inch) per day.**

- **Fingernails tend to grow a little faster than toenails.**

- **Toenails are approximately twice as thick as fingernails.**

- **In general, nails tend to grow faster in summer than they do in winter.**

- **Men's nails usually grow faster than women's nails.**

- **Nails on your dominant hand tend to grow faster.**

CHAPTER 3

The party continues and has taken on a much more serene and romantic tone. Leyner is on the couch with his Cinderella and is eating cocktail egg rolls off her webbed toes as they share his bottle of tequila. I am finally free from the body questions and at last can enjoy a drink myself. It

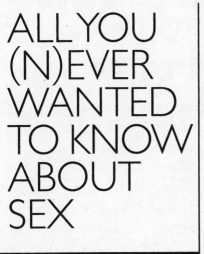

ALL YOU (N)EVER WANTED TO KNOW ABOUT SEX

seems as though the drama of the evening is over until I hear Leyner shout, "Can anyone get me some duck sauce?" Leyner is too impatient and his libido is obviously inflamed, so he is unable to wait for the hired help to procure the requested Chinese condiment.

I cross the crowded room and find Leyner conducting a focus group on homemade and alternative lubricants. As I reach the front of the group, Leyner is now meticulously mixing exact proportions of Tabasco, runny Brie, and a dash of his sacred tequila to form what he now calls his "spicy sex balm." I try to suggest that the Tabasco may cause contact dermatitis on the more sensitive parts of the body, but Leyner chooses not to heed my medical advice as he leads his new "special" friend to his ad hoc laboratory.

I am left behind to answer a sudden barrage of questions about sex. With anatomically correct dolls, *Sex and the City*, and Internet porn, you'd think there'd be nothing left to learn. But there are still some questions people are afraid to ask until they've had that third martini.

IS SPERM NUTRITIOUS? OR FATTENING?

You are what you eat. In this case, it is somewhat true, as sperm contains important genetic material. But sperm, despite its important load, is not particularly nutritious or fattening. The average ejaculate, about one teaspoon, contains between two and three hundred million sperm. Total calories: about five. These calories are derived from protein, including enzymes and sugars (mainly fructose) secreted into semen by the prostate gland to provide the sperm with the energy to swim.

Other good stuff found in semen includes water, vitamin C, citric acid, phosphate, bicarbonates, zinc, and prostaglandins. A veritable breakfast of champions.

Gberg: I still can't believe we ended up with this title.

Gberg: I still miss "Cocktail Party Medicine."

Leyner: I know . . . I can't even look at my own nipples anymore without blushing with shame.

Gberg: Or if you had your way, it would be "Is Sperm Fattening?" That would be even worse.

Gberg: It hurts the neck to stare at your nipples for too long.

Leyner: I still love that title.

11:50 A.M.

Gberg: That would have meant the end of my medical career.

Leyner: To the contrary—I think it would have landed you a Surgeon General nomination in the Hillary Clinton administration.

CAN YOU GET PREGNANT WHILE YOU ARE HAVING YOUR PERIOD?

In the 1970s there was an after-school special on TV where two girls were talking about whether you could get pregnant from kissing a boy while wearing a wet bathing suit. That definitely isn't true, but the answer to this question is a little more complicated.

The basic answer is yes, you can.

First, not all bleeding is a real period. Sometimes a woman can have spotting during ovulation and that would be a very easy time to get pregnant.

Also, if a woman has a short menstrual cycle (about twenty-one days), then she could be ovulating on day seven of her cycle. This would be the seventh day from the first day of her period, and if her period lasts seven days, then it is possible that her period is ending just as she is ovulating.

Confusing? It is definitely easier to practice safe sex.

DO OYSTERS REALLY MAKE YOU HORNY?

Oysters definitely do resemble a certain anatomical part, but this doesn't make them stimulants. There are a lot of different foods and substances that are thought to be aphrodisiacs. Though there is no science to prove that any of them work, there are some theories on oysters.

Oysters are full of many vitamins and minerals, especially zinc. Zinc controls progesterone levels, which have a positive effect on the libido. Other foods and products that are thought to get your mojo working include:

chocolate

strawberries

champagne

Spanish Fly

animal crackers (but only if consumed during
   sex and dunked in Spanish fly)

BIG HANDS, BIG ____?

When I was in Spain, I learned that the Spaniards believed that the distance measured from either side of your nose across the tip to the other side hinted at your penis size. As the owner of a rather large schnoz, I, Billy, was happy about this. But, by this same measurement, Geppetto must have had quite a task making pants to hold Pinocchio, and Cyrano de Bergerac wouldn't have had to waste so much time writing love letters for others. Unfortunately, none of this is true.

There also seems to be no evidence that hand size is related to the size of your package.

When it comes to foot size, two urologists, in a study in the *British Journal of Urology,* measured the stretched penile length of 104 men and related this to their shoe size. They found that there was no relationship.

As for other size issues, a small penis expands more than a big one during erection. And since a woman's sexually sensitive parts are in and around the outer third of the vagina, a big penis is not necessarily needed to satisfy her. Size doesn't *really* matter, ladies.

WHY DO MEN WAKE UP IN THE MORNING WITH AN ERECTION?

You must be referring to nocturnal tumescence! Or do you mean Private Willie Is Playing Reveille, or perhaps Morning Glory? There are many names for this morning starter, but only one good explanation for its presence.

These erections are experienced in the REM (rapid eye movement) phase of sleep. REM sleep is more frequent just before waking up.

Erections can also happen at other times without any stimulus. There is no scientific reason why these spontaneous erections always seem to happen in the most embarrassing places (parties, holidays, movies, weddings, funerals).

Professor Leyner has written extensively on the subject. The noted scientific research journal *Maximum Golf* contained this exerpt:

••

Announcer B: If you're just joining us, Michael Neubauer is still in bed—but we've got an interesting development here.

Announcer A: That's an erection, Bobby. That's a classic morning erection. You couldn't diagram it any better than that. The corpora cavernosa and the corpus spongiosum are engorged with blood and rigid. Absolutely paradigmatic. What do you think?

Announcer B: I think you just have to give it a good whack here, Chris.

Announcer A: Apparently you've got the ear of Michael Neubauer, because that appears to be precisely what he's going to do.

Announcer B: I really like what he's trying right now. He's running this frenetic montage of actresses, models, and athletes through his mind: We've got Renée Zellweger, Neve Campbell, Liv Tyler, Catherine Zeta-Jones, Britney Spears, Salma Hayek, Foxy Brown, Niki Taylor, Lil' Kim, Melissa Joan Hart, Charlize Theron, Sarah Michelle Gellar, Zaha Hadid, Benazir Bhutto, Se Ri Pak, Karrie Webb, Serena Williams; Anna Kournikova, Jelena Dokic, Mary Pierce. Now he's going way back: we've got some high school yearbook action; we've got some baby-sitter action. He's just totally scouring the memory bank here—there's one of his grandmother's mah-jongg partners with the arm wattles. . . .

Announcer A: What are you trying to do in this situation?

Announcer B: You're trying to get some erotic traction. You're looking for that face or that body that just clicks, y'know, that you can fixate on, and then you try to get the voice, the incantatory exhortations, "Michael . . . oh yeah . . . oh Michael . . . oh my God!"

Announcer A: How do you like his mechanics right now, Bobby?

Announcer B: Excellent. He's got a very good rhythm going right now.

Announcer A: Marvelous touch. You sense he has every-thing to gain and nothing to lose.

Announcer B: One caution here. You want to avoid the computer and the books over there—

Announcer A: I think we're reaching critical mass, Bobby.

Announcer B: They don't come much better than that, Chris. Absolutely marvelous shot!

• •

CAN PEOPLE IN WHEELCHAIRS STILL HAVE SEX?

If an aroused male is unfortunately the victim of a spinal cord injury, the ability to have sexual intercourse depends on the level of injury. In men there are normally two types of erections, psychogenic erections, which result from sexual thoughts, and reflex erections, which result from direct physical contact.

Psychogenic erections develop from the nerves of the spinal cord that exit toward the bottom of the spine at the T10-L2 levels. Generally, men with an incomplete injury at a low level are more likely to have psychogenic erections than men with high-level incomplete injury. Men with complete injuries are less likely to experience psychogenic erections.

Reflex erections arise in the sacral area of the spinal cord. Many men with a spinal cord injury are able to have a reflex erection with physical stimulation if this pathway is not damaged.

IS IT TRUE THAT YOU CAN BREAK YOUR PENIS?

It pains us to say, but this is true. There is no bone in your "bone," but you can rupture the penis, which is called a penile fracture. Sudden trauma or bending of the penis in an erect state can break the thick fibrous coat surrounding the corpora cavernosum tissue that produces an erection.

This happens most frequently during sex. Don't expect to get a cast and crutches, though. This injury is an emergency and requires surgery to prevent sexual dysfunction. Ouch!

DOES MASTURBATION CAUSE STUTTERING, BLINDNESS, OR HAIRY PALMS?

As a doctor, there is a great deal of medical research to read. Piles of *The New England Journal of Medicine* and the *Annals of Emergency Medicine* surround the house. Rarely does this drudgery cause excitement, unless you stumble across the April 7, 2004, issue of *The Journal of the American Medical Association*.

As a young boy you get bombarded by the fear that masturbation can cause stuttering, blindness, or hairy palms. Even if you escape these maladies you can still be left feeling like a degenerate if you engage in self-love too often. Fear no more. In the article "Ejaculation Frequency and Prostate Cancer" you finally find out that whacking off may be good for you. This article states that ejaculation frequency is not related to increased risk of prostate cancer and that the group with an ejaculation frequency of greater than twenty-one times per month had a lower risk of prostate cancer. Great! And *no* hairy palms.

DOES USING A TAMPON FOR THE FIRST TIME MAKE YOU LOSE YOUR VIRGINITY?

Dr. Billy's young cousin's friend asked him this one, another common question. There are many different opinions about what constitutes losing one's virginity, but most experts agree that women lose their virginity the first time they have vaginal intercourse.

This question mainly revolves around the idea that only an intact hymen (a fold of tissue that partly covers the entrance to the vagina) proves that a woman is a virgin. It is possible for a woman's hymen to become stretched open during activities like playing sports or masturbating. Hymens have at least one opening that will allow menstrual flow out of the body. Tampons may stretch the hymen a little bit, but they don't usually stretch it open all the way. Either way, most people in the modern age believe that the condition of the hymen does not have anything to do with the definition of a virgin.

WHY DO NIPPLES BECOME ERECT?

Small muscle cells arranged cylindrically within the nipple are responsible for the nipple becoming erect. This occurs when they are stimulated by cold temperature or sexual arousal.

WHAT CAUSES SHRINKAGE?

A penis whose flaccid length when stretched is more than approximately 2.5 standard deviations below average size for his age group is referred to medically as a micropenis. This is not a title that any self-respecting man wants to put in his bio, but perceptions of penis size can be deceiving.

Cold air, cold water, fear, anger, or anxiety can cause the penis, scrotum, and testicles to be pulled closer to the body, thereby shortening it to micropenis lengths.

Warm conditions on the other hand can cause the penis to lengthen. Although the size of the nonerect penis differs widely from one male to another, this variation is less apparent in the erect state. Even *Seinfeld* weighed in on the shrinkage issue.

Some penis reference points:

1. *The Kinsey Report,* 1948: average length of 6.20 inches (15.25 cm) with a standard deviation of .77 inches (1.96 cm).
2. Study by Wessells et. al., 1996: average length of 5.1 inches (13.0 cm).
3. Other studies: average length of 5.7 inches (14.5 cm).

DOES CIRCUMCISION LESSEN THE FUN OF SEX?

The debate over circumcision has been going back and forth for years. Some American medical experts believe that all newborn males should be circumcised. The side in favor cites the decreased rate of urinary tract infections and sexually transmitted diseases among circumcised men, and the other emphasizes the pain and stress inflicted on a baby during circumcision. Others believe that the benefits of circumcision are not great enough to justify the possible complications of the procedure. In 1999, the American Academy of Pediatrics Task Force on Circumcision found potential medical benefits of circumcision, but decided that the evidence was not strong enough to recommend routine circumcision. But, more important, back to sex.

There are several studies found in urology literature that look at the effect of male circumcision on male sexual satisfaction. These studies found conflicting answers.

The effect of male circumcision on the sexual enjoyment of the female partner was also examined in a study in the *British Journal of Urology*. The authors concluded that "women preferred vaginal intercourse with an anatomically complete penis over that with a circumcised penis." Interestingly, the authors of the study also wrote and published a book entitled *Sex As Nature Intended It*. They also recruited some of their volunteers for the study from an announcement in an anticircumcision newsletter. We mention this not as circumcised males but as impartial interpreters of the medical literature.

Interesting circumcision facts:

- -

The twelfth-century physician and rabbi Moses Maimonides advocated male circumcision for its ability to curb a man's sexual appetite.

Male circumcision was introduced into English-speaking countries in the late 1800s as a method of treating and preventing masturbation.

Male circumcision, the most commonly performed surgery in the United States, removes 33 to 50 percent of the penile skin, as well as nearly all of the penile fine-touch neuroreceptors.

- -

DOES THE KIND OF UNDERPANTS MEN WEAR AFFECT THEIR FERTILITY?

Everyone has a preference.

President Clinton revealed on MTV that he preferred boxers.

Kramer on *Seinfeld* said, "I need the secured packaging of jockeys. My boys need a house!"

The question is whether or not there is any science behind making this decision.

It was originally thought that wearing tight underwear could lead to infertility in men as it may raise testes temperature to a point where it interferes with sperm production.

In 1998 in *The Journal of Urology*, Drs. Robert Munkelwitz and Bruce R. Gilbert analyzed semen samples from ninety-seven men with fertility problems. Half of them wore briefs, the other half wore boxers. The researchers measured the men's scrotal, internal, and skin temperatures, both while the men were wearing underwear and again when they were going commando. They found no significant differences between the two groups of men in scrotal temperature, sperm count, sperm concentration, or sperm motility.

The purpose of the scrotum is to maintain the testes at a temperature of approximately five degrees less than the rest of the body (about 93.6 degrees F). It appears as though the scrotum does its work whether you are in boxers, briefs, or free ballin'.

So you can make your own fashion choice and your sperm won't know the difference.

IS THERE REALLY A G-SPOT?

An article in the *American Journal of Obstetrics and Gynecology* in 2001 called the G-spot a sort of gynecologic UFO. Well, the authors of that article may want to start readying themselves for space travel. That is, after they read this paper from Cairo University: "The Electrovaginogram: Study of the Vaginal Electric Activity and Its Role in the Sexual Act and Disorders." In this paper, the authors investigated the hypothesis that the vagina generates electric waves, which affect vaginal contraction during penile thrusting. They found electric waves could be recorded from the vagina.

They also postulated that there was a vaginal pacemaker that seems to represent the G-spot, which is claimed to be a small area of erotic sensitivity in the vagina.

So what is this vaginal Loch Ness Monster?

The G-spot is simply a small area located on the upper wall of the vagina, toward the belly, about two to three inches from the vaginal opening. The G-spot was named in honor of Ernst Grafenberg, a German physician who, in the 1950s, wrote an article that mentioned an erotic zone on the anterior wall of the vagina that would swell during sexual stimulation.

There are various opinions on the best way to find the G-spot. Some women say that being on top during intercourse works best. Others swear by rear-entry as the best way to hit the G-spot. Some even say that because of its location, a shorter, smaller penis may actually be more effective at reaching the G-spot. A clue to its location may be that some women feel a sudden urge to urinate when their G-spot is touched—not surprising since the G-spot is located right near the urethra.

DO KEGEL EXERCISES REALLY WORK?

For those of you who have never heard of Kegel exercises, don't expect to see a class offered at your gym.

Kegel exercises were originally developed as a method of controlling incontinence in women following childbirth. They're named after Arnold Kegel, the Los Angeles doctor who promoted their development in the 1940s. The principle behind Kegel exercises is to strengthen the muscles of the pelvic floor, or the pubococcygeus muscles. These muscles run from the back to the front of your pubic bone and encircle the openings of the vagina and rectum. Strengthening them helps improve the urethra and rectal sphincter function.

These exercises are recommended for women with urinary stress incontinence, but many others do Kegel exercises for more fun reasons.

Advocates believe that there are several benefits for a woman who exercises her vagina. They claim that it makes it easier for her to reach orgasm, makes orgasms stronger or better, and makes the vagina more sensitive.

CAN HOT TUBS MAKE YOU INFERTILE?

Heat is damaging to the sperm, and theoretically can affect male fertility to a certain extent. But, there is no clear scientific evidence that implicates hot tubs. There may be a temporary reduction in sperm function after a soak, and prolonged repetitive use could cause problems, but none severe enough to avoid the occasional dip.

Saunas do not appear to influence fertility either. A few studies have reported decreased sperm count or decreased sperm movement after sauna use, but in Finland, where saunas are most common, men have high sperm counts and no apparent fertility issues.

DO MEN NEED SEX MORE OFTEN THAN WOMEN?

Men are often told that they have sex on the brain, and it appears as though this may be true.

In one recent study in *Nature Neuroscience,* a team of researchers had twenty-eight men and women look at erotic photographs while an MRI took scans of their brain. The subjects looked at arousing photographs of heterosexual couples engaged in sexual activity, sexually attractive nudes of the opposite sex, and at pictures of men and women in nonsexual situations. When analyzing the MRI results, researchers found that two areas of the brain, the amygdala and the hypothalamus, were more activated in men than in women when viewing identical sexual stimuli.

So, do men have sex on the brain? Of course, and do bears shit in the woods?

We really didn't need a team of researchers to answer this one.

CAN A MAN EVER RUN OUT OF SPERM?

What a nightmare. The well running dry. Could it be possible?

The answer is no, but there is some bad news.

A woman is born with all the eggs she will ever have, but a man's supply of sperm is renewed throughout his life. This sounds great, but researchers have found that men over thirty-five have more abnormalities in sperm movement and sperm with more seriously damaged DNA than younger men. It also has been found that over time, average sperm counts have been decreasing. The World Health Organization guidelines say a normal sperm count consists of twenty million sperm per ejaculate, with 50 percent motility and 60 percent normal morphology. This is different than twenty-five years ago when the normal count was near one hundred million.

ARE THERE ANY SPECIFIC THINGS THAT AFFECT THE SCENT OF A WOMAN?

My wife's "friend" once told her that eating pineapple made you smell good "down there." The friend had heard this from a call girl. It doesn't get more evidence based than that.

There is, however, no scientific research on this sensitive subject. People do believe that you are what you eat, so what you ingest, ladies, can affect the smell and taste of your womanly secretions. Foods that are often mentioned as having the potential to cause problems "down there" are asparagus, garlic, and curry.

CAN A WOMAN EJACULATE?

There have been many claims about female ejaculation, but this was always dismissed as urination during intercourse. More recent evidence has found that higher levels of a compound, prostatic acid phosphatase, has been found in patients who claim to have female ejaculation. This compound is also found in high levels in male ejaculate and originates in the prostate.

Researchers have taken an anatomic approach to the issue of prostatelike components in female ejaculate. They believe that if women ejaculate a fluid that is not urine, then it must be coming from someplace other than the bladder. The most likely source was thought to be the female paraurethral glands or Skene's glands.

Autopsy tests of Skene's glands have found substances identical to those found in the prostate. Some experts now call these glands the female prostate. So, it seems highly likely that some women can, in fact, ejaculate, but causing that to happen may be as tough as finding the G-spot.

WHAT IS A HICKEY?

A hickey is a bruise that forms when a person sucks and lightly bites an area on another person's body, causing the blood vessels under the skin to break. It is also a badge of honor for horny fifteen-year-olds all around the world.

CHAPTER 4

I look at my watch and I can't believe that it is only 12 A.M. I answer the last of the sex questions and quickly scan the room to make sure that my path is clear to the bathroom. I see an opening and rush off hoping to avoid any more questions.

CAN I TREAT IT MYSELF?

The door is slightly ajar and I push it open tentatively to find our hostess, Eloise, sitting on the edge of the bathtub. Leyner is clutching the massaging showerhead and is directing a cold stream of water on her burned and blistered cheeks. As it turns out, Eloise, ever the impeccable hostess, had joined Leyner and Cinderella in their laboratory to see if they needed their glasses refilled. Leyner then insisted on slathering her with his spicy balm. Little did he know that Eloise had had a deep-cleansing facial peel just a few hours before the party, leaving her skin ultra-vulnerable to any corrosive ointments. It was comforting for me to see that in the heedless anarchy of the moment, Leyner had actually done the right thing and was correctly caring for her burns.

There are two venerable maxims that rule the professional classes. For doctors it is do no harm, and in the world of jurisprudence it is anyone who represents himself in court has a fool for a client. So, if you must be your own physician, do no harm and don't be a fool.

CAN YOU TAKE THE TETRACYCLINE MEANT FOR FISH TANKS?

A patient came to the ER one day for treatment of an infection on her scalp. She had tried to treat it at home by taking the tetracycline for her fish tank. As it turns out, she actually didn't need antibiotics for her rash. Was this self-medicating ingenuity or insanity on her part?

I wouldn't exactly call it ingenuity, because the other great idea she came up with was to rub toothpaste on her head. As for the fish tank tetracycline, it seems as though those tablets often contain the same dosage as your over-the-counter tablets, but I certainly cannot vouch for their purity. Logic says to stick with the stuff you get from the pharmacy.

DOES CRANBERRY JUICE CURE URINARY TRACT INFECTIONS?

Maybe. A study in *JAMA*, or *The Journal of the American Medical Association* for all not in the know, demonstrated a significant reduction in the rate of urinary tract infections in older people who drank cranberry juice daily. We don't know exactly why, but the most likely answer is that some chemical in the juice prevents bacteria from sticking to the wall of the bladder. I know some people don't like cranberry juice, but since urinary tract infections can happen after sex, it's probably worth it to drink a little bitter cranberry juice instead of smoking a cigarette.

WILL YOGURT CURE A YEAST INFECTION IF YOU PUT IT "INSIDE"?

Yogurt may have some role in helping prevent yeast infections, but only when you put it in the correct orifice—your mouth.

Yogurt does have some very interesting health properties. Some women find eating one cup of yogurt a day while taking antibiotics is helpful in preventing the yeast infections that often follow such treatment; however, yogurt alone will not cure vaginal yeast infections that are already in full bloom.

DOES CANDLE FLAME REMOVE EARWAX?

Excessive amounts of earwax can cause decreased hearing or pain, but this is no reason to start lighting your head on fire. "Ear candling," or coning, involves placing a cone-shaped device in the ear canal and, with the help of smoke or a burning wick, removing earwax. The companies that make these devices claim many other health benefits. The FDA and the Canadian government disagree, and both have come out against these fraudulent claims. In most cases earwax will come out on its own, and if not, you should see your doctor. Avoid these silly products and save the cones for your ice cream.

105 CAN I TREAT IT MYSELF?

IS IT SMART TO PUT BUTTER ON A BURN?

No. Save the butter for breakfast.

Butter is the wrong thing to put on a burn. It will trap the heat in the skin and prolong the pain. Use cool water instead.

I do have to say that there are some kitchen supplies that may be useful for the pain of a burn, straight from the medical literature in India: boiled potato skins and honey.

Why honey? Honey is sometimes used for its antibacterial effects. Boiled potato skins may seem to be an unusual treatment, but they maintain a moist environment.

The best thing to do when you grab a hot pan is to cool your burn under running water, and only after it has thoroughly cooled, apply an antibiotic ointment. For severe burns you should go to your local ER.

DOES MELATONIN WORK FOR JET LAG?

Here is one for all the world travelers. Melatonin may be an effective solution for your problems with jet lag.

A review of ten studies on the use of melatonin concludes that two to five milligrams of melatonin taken at bedtime after arrival at your destination is effective and may be worth repeating for the next two to four days. This should be done in conjunction with nondrug measures—such as avoiding dehydration and alcohol, engaging in exercise and activity during daylight hours, eating well, sleeping well, and adjusting to the local time schedule—for fighting the dreaded jet lag.

SHOULD YOU PUT A STEAK ON A BLACK EYE?

Ice is hardly as dramatic as a carefully placed porterhouse, but it does the same job. There is no magic in the beef, just cold and a little pressure. Keeping your head elevated and avoiding aspirin or ibuprofen, which can affect the ability of your blood to clot, also helps. The best idea is to avoid getting punched in the first place.

WILL TOOTHPASTE GET RID OF ZITS?

There are many home remedies for skin ailments, and I've seen patients come to the emergency room covered in all sorts of creams and potions. Some common antizit home remedies people try concocting include baking soda, vinegar, coffee grounds, Mercurochrome (a red substance no longer sold in this country), iodine, hemorrhoid cream, sugar, salt, and toothpaste.

It is commonly believed that toothpaste on zits is an excellent home remedy. There are no scientific studies that I could find on the use of toothpaste for acne but it may work to dry out those troublesome blemishes.

But if you go the toothpaste route, there are some things to look out for. Perioral dermatitis is an eruption of discrete papules and pustules on an erythematous scaling base around the mouth (fancy description for "acnelike"). It occurs almost exclusively in women between the ages of twenty and thirty-five. The cause is unknown, but some people think that fluorinated toothpaste may be a factor.

Fragrance allergies are also a danger with the toothpaste method. Balsam of Peru is an ingredient that has been known to cause allergic reactions and cinnamic aldehyde in toothpastes has also been a common culprit. We say to stick with Clearasil.

IS IT DANGEROUS TO POP ZITS?

In an unrelated but truly bizarre story, one night I was working in the ER and a patient came in with a bandage over her jaw. I asked her what happened and she was very timid in responding. She said she was embarrassed because she had picked her face over and over, which had caused an infection. I tried to calm her, told her that this was common, and asked her if I could take a look. When she removed her bandage, she revealed a 4 by 4-inch hole all the way down to her jawbone. She received some antibiotics and a careful psychiatric evaluation and at no point did anyone say, "Didn't your mother tell you not to pick at your face?"

As for zit popping, it definitely can lead to some complications. Squeezing pimples can actually push the zit-causing bacteria farther into the skin, causing more redness and swelling. It is also the most common cause of acne scarring.

There is one more deadly complication from zit popping, which is called cavernous sinus thrombosis, a blood clot in the sinus cavity that runs between the sphenoid bone, the large bone at the base of the skull, and temporal bone located near the temple. The real danger zone for zit popping is an area that some people refer to as the triangle

WHY DO MEN HAVE

of death, an area stretching from the bridge of the nose to the corner of the mouth to the width of the jaw. The veins in this area drain into the cavernous sinus and any severe infection in this area can cause cavernous sinus thrombosis. Squeezing zits in this part of your face can cause an infection and start this dangerous process.

IF SOMEONE IS CHOKING AT A DINNER PARTY, CAN YOU DO A TRACHEOSTOMY WITH AN OYSTER KNIFE?

Our friend Kim can do pretty much anything. She was like Martha Stewart before anyone had seen her bake her first cookie. Add to that a rugged edge that allows her to take on any project. She wanted to know the answer to this one, and we realized if anyone could do this, it would be her. She also wouldn't allow the procedure to interrupt her dinner party.

A cricothyroidotomy (similar to a tracheostomy) is one of the most dramatic procedures done in the emergency room. This procedure is an emergency attempt to relieve a blocked airway. Remember the *M*A*S*H* episode where Father Mulcahy sticks a pen into some guys throat to help him breathe?

The oyster knife might work too, but definitely avoid trying this at home! Call 911 instead.

DOES URINATING ON A JELLYFISH STING STOP THE BURN?

We all saw that *Friends* episode (c'mon, you watch it, too) when Monica gets stung by a jellyfish. Joey remembers that peeing on a jellyfish sting takes the pain away, Monica "can't bend that way," and Joey gets "stage fright," leaving Chandler to save the day. Don't believe everything you see on TV.

Most jellyfish stings cause only pain and numbness. The Australian box jellyfish is the most venomous and deadly of all stinging marine creatures. Approximately 20 percent of those stung by the box jellyfish will die. Portuguese man-of-war is also dangerous but nothing compared to the box jellyfish.

The following guideline can be applied to most jellyfish stings: The patient should remove any visible tentacles, using gloves if possible. The area of the sting should be rinsed with household vinegar. The acetic acid of the vinegar can block discharge of the remaining nematocysts (stinging cells) on the skin and should be applied liberally. If vinegar is not available, salt water can be used to wash off the nematocysts.

In laboratory tests, urine, ammonia, and alcohol can cause active stinging cells to fire, which means applying them has the potential to make a minor sting worse, so urinating on a jellyfish sting is both gross and painful.

WHY IS IT BAD TO INSERT COTTON SWABS IN YOUR EARS?

Oh the pleasure of the forbidden! Those things that you are not supposed to do are always so enticing.

The ears, for the most part, do not require any routine cleaning. Ears are like a self-cleaning oven. With the help of gravity and body heat, earwax will gradually find its way out. If wax appears on the outer ear, a cotton swab may be used. If you can't help but go in farther, you are risking wax impaction or injury. If you do get wax impacted in your ear, you will be in pain and half deaf. There are over-the-counter preparations that can help relieve wax blockage but warm water in a syringe often works. As a last resort you can see an ear doctor or come to the ER for a good cleaning.

It is not uncommon for us to see patients who have violated these rules and come to see us to remove the tip of the cotton swab that has fallen off inside the ear. Don't worry, we are prepared. We also remove other things like cockroaches, beads, and pen caps, all of which we've pulled out of ears.

• •

Gberg: We need a list of things for the cotton swabs in the ear question.
Leyner: OK.
Gberg: Things that you aren't supposed to do but can't resist.

6:05 P.M.
Leyner: Picking scabs.

Gberg: I love it when they bring the food to the table and say "hot plate."

Gberg: Can't help but touch.

Leyner: That's good!

Leyner: More . . .

Gberg: Like Carrie says, "More funny, boys."

Gberg: Making fun of the editor is one of those things that you are not supposed to do but can't resist.

Leyner: More funny like, "How do you extract my size 9 old school Adidas shell toe from the rectum of a book editor?" More funny . . .

Gberg: Should I leave that in?

6:10 P.M.

Leyner: Your call.

Leyner: Might be a little harsh.

Leyner: But it's from the heart.

Gberg: Slightly.

Gberg: A dangerous little muscle, that heart of yours.

Leyner: I'm trying to think of more not-to-do stuff.

Gberg: Talk at a woman's breasts.

Gberg: Eat your young.

Leyner: Pick chicken pox.

Gberg: You just want to pick stuff.

Leyner: Pop pimples.

Gberg: Talk with food in your mouth.
Leyner: I knew some girls who loved popping each other's and their boyfriends'
pimples.

6:30 P.M.

Gberg: I wish my lady would groom me like
a monkey.
Leyner: It's all in the eating . . . monkeys
combine grooming and eating . . . that's the
special part . . . picking insects out of
our fur and eating them.
Leyner: Metaphorically speaking.
Gberg: You are at your best when speaking
metaphorically.
Leyner: Thank you again.
Gberg: I gotta leave to go to the Knicks
game soon.
Leyner: Can we get back at this tomorrow
when you get back from the hospital?
Gberg: Let's try to finish everything.
Leyner: You have to go, right . . . we'll
talk about it tomorrow.
Gberg: OK, let's talk tomorrow.
Leyner: I'll look at the e-mail . . . and
we'll drive the final stake into the heart
of this vampiress tomorrow.

· ·

IS IT DANGEROUS TO PERFORM COLONIC IRRIGATION ON YOURSELF?

Colonic irrigation claims to help indigestion and yeast infections, control blood pressure, restore pH balance, reduce bad odors, clear colon blockage, induce proper blood clotting, stimulate production of white blood cells, help prevent gallstone production, clean the colon of parasites, help loss of concentration, and aid lung congestion, sinus congestion, skin problems, and nail fungus.

Not a bad day's work, but not exactly proven, and yes, potentially dangerous.

Colonic irrigation (CI) is a procedure in which very large quantities of liquids are infused into the large intestine, or the colon, via the rectum through a tube. The purpose is to detoxify the body through the removal of accumulated waste from the colon. This may involve the use of twenty or more gallons of liquid. Liquids used in colonics may contain coffee, herbs, enzymes, or wheatgrass.

The machines used for colon therapy are illegal unless used during conventional medical treatment. Colon therapy also can be dangerous. Complications include bowel perforation, heart failure from excessive fluid absorption, electrolyte imbalance, and several outbreaks of serious infections. One case linked to contaminated equipment caused amebiasis, a parasitic infection, in thirty-six people.

• •

11:50 A.M.

Leyner: This book is going to ruin both of us. The editor will probably get a huge promotion and we'll end up in the subway wearing fedoras and playing Andean flute music.

Gberg: Andean flute music sounds appropriate for the theme music in a colonic ad.

11:55 A.M.

Leyner: Why are people so interested in colonics?

Gberg: Seems insane to me.

Gberg: Who says it's supposed to be clean, that is, your colon?

Gberg: I sounded like Yoda there.

Leyner: Me too . . . part of the wonderful Judeo-Christian legacy of self-loathing . . . you know . . . how we're essentially filthy inside.

Gberg: Putrefaction.

Gberg: I feel like I am rotting inside today.

Leyner: That's right . . . If you can't have a dirty colon . . . c'mon.

Gberg: Good title for a pop song.

Leyner: Sounds good for Prince.

Gberg: Imagine Britney Spears singing the Mark Leyner version of "If you can't have a dirty colon . . . c'mon."

Gberg: Great video too.

Leyner: I love tracking the doings of aging rock stars.

Gberg: Some celebrity colonoscopy cameos.

12:00 P.M.

Leyner: Rod Stewart's become the new Jim Nabors somehow.

Gberg: What is he up to now?

Gberg: I read something on Page Six about Elton John injecting himself with lamb's urine to lose weight.

Leyner: Rod's singing duets with Dolly Parton, Gershwin ballads, and children's songs . . . and doing an album with the Wiggles next—wouldn't surprise me.

Gberg: They also said "Michael Jackson reportedly used to keep his weight down with lots of self-administered enemas, but later needed a tampon to control 'embarrassing leakage.'"

Leyner: How do you get a lamb to pee in a cup?

Gberg: That is going to be my next job after the book—catheterizing lambs.

Leyner: Jackson should let himself get fat like Elvis did.

Leyner: Is that true about Elton John and lamb urine?

Gberg: Page Six, my friend. Check the New York Post. Isn't everything in the newspaper true?

Leyner: Yes . . .

Gberg: Maybe we can add the lamb's urine question in as an urban legend.

Leyner: Doesn't something have to be in the public subconsciousness for more than a day to qualify as an "urban legend"?

• •

DOES BREAST MILK CURE WARTS?

Here's one from the June 2004 *New England Journal of Medicine*. A cream containing an ingredient of human breast milk appears to be an effective treatment for stubborn warts. The key ingredient of the cream is a compound called alpha-lactalbumin-oleic acid. Its Swedish creators have nicknamed the cream HAMLET, for Human Alpha-lactalbumin Made Lethal to Tumor cells.

This may lead to other areas of research as certain types of warts or human papilloma virus (HPV) can be linked to cervical cancer.

No word yet on whether Starbucks will be introducing a tall-decaf-breast latte.

IF YOU GET BITTEN BY A SNAKE, SHOULD YOU SUCK OUT THE VENOM?

I love a good Western and nothing could be more bad-ass than biting into a snake wound and spitting out the venom. Of course this would be followed by some whiskey and a good gunfight.

Unfortunately, this is no longer an accepted practice. Sucking at a snakebite is not only ineffective but could lead to an infection at the wound site.

According to the American Red Cross, these steps should be taken after a snakebite:

1. Wash the bite with soap and water.
2. Immobilize the bitten area and keep it lower than the heart.
3. Get medical help.

Toxicology experts might also suggest applying a tourni-quet loosely above the bite to prevent the venom from spreading. This must be done with caution, as the tourni-quet itself can cause problems if it cuts off the blood flow entirely.

The person then needs to be transported rapidly to an emergency room. Antivenin is available for a variety of different snakes. Other treatments include antibiotics and surgery.

Of the estimated one hundred and twenty different types of snakes found in the United States, about twenty are poisonous. Most bites occur in the southwestern part of the nation, but they even occur in New York City. In New York State there are three species of poisonous snakes, the timber rattlesnake, the massasauga rattlesnake, and the copperhead. In the city, however, most bites occur from snakes that are kept as pets.

WHAT ARE HICCUPS, AND HOW DO YOU GET RID OF THEM?

Doctors are known for using complicated words that make them sound either extremely intelligent or really out of touch with what most people can understand. The medical word for hiccups, singultus, is a perfect example of when physicians sound ridiculous.

Hiccups are caused when the diaphragm becomes irri-tated and pushes air rapidly up in such a way that it makes an irregular sound.

Some things that irritate the diaphragm and cause hiccups are distention of the stomach from food, alcohol, or air, sudden changes in gastric temperature, or use of alcohol and/or tobacco in excess. Hiccups also can be caused by excitement or stress.

While most cases of the hiccups last only a few minutes, some cases of the hiccups can last for days or weeks. This is very unusual, though, and it's usually a sign of another medical problem, such as injections near the diaphragm, hiatal hernias, severe gastroesophageal reflux disease (GERD), or a tumor irritating the nerves in the chest. Hiccups lasting longer than one month are termed intractable or incurable. The longest recorded attack of hiccups is six decades. Doctors sometimes use the antipsychotic drug Thorazine to treat intractable hiccups.

If you don't want to go the antipsychotic route, you could try one of these simpler but unproven cures:

1. Breathing into a paper bag.
2. Drinking out of a cup from the side opposite your mouth.
3. Holding your breath.
4. Eating a teaspoon of sugar.
5. Sucking on a wedge of lime or lemon.
6. Drinking a glass of water with a straw while you plug your ears with your fingers.
7. Pulling the top of your hair for one to two minutes.
8. Placing a cotton swab in the roof of your mouth and gently rubbing.
9. Pulling hard on your tongue.

DOES BATHING IN TOMATO JUICE REMOVE THE SMELL OF A SKUNK?

For those of you who were watching TV in 1970, you may have seen episode 8 of the first season of *The Partridge Family* when a skunk finds its way onto the family bus and turns the Partridges into stinkers. Reuben remembers that tomato juice can remove the skunk odor, so the family bathes in it. All is well until the family dog gets them covered again. Without time to take another tomato bath, the band plays their concert at a children's hospital from inside a glass-enclosed operating room. That's great TV.

The major molecules that make skunk spray smell are sulfur compounds. It is a common belief that tomato juice removes the smell, but there is no scientific evidence to support this claim. The tomato juice probably just tricks the nose into not recognizing the skunk smell through the overpowering red gravy scent. One recommended treatment for pets is one quart 3-percent hydrogen peroxide, one cup baking soda, and one teaspoon mild dishwashing detergent. People can try the same, but be careful; the peroxide can have a bleaching effect.

DOES EATING FRESH PARSLEY CURE BAD BREATH?

Parsley was used in the past in medicinal recipes for cure-alls, general tonics, poison antidotes, and kidney and bladder stone relief. Parsley is rich in vitamins and minerals, particularly vitamins A and C. It is also said to be rich in antioxidants. Parsley also can relieve bad breath, although there are no medical studies linking it to halitosis. It is good to note that there are two varieties of parsley: curly-leafed and flat-leafed, which has the stronger flavor. Therefore, the flat-leaf kind is better to cure your chili dog breath.

DOES WARM MILK REALLY HELP YOU SLEEP?

There doesn't appear to be a great deal of research on the role of milk as a sleep aid. Milk is certainly a simpler alternative to prescription sleeping pills or drinking so much you just pass out, and there are several theories as to why it might work. Milk contains tryptophan, the same ingredient that makes everyone sleepy after Thanksgiving dinner (see food coma question on page 44). The warmth of the milk can have a minimal effect on your body temperature and sometimes make sleep a little easier. Milk also contains melatonin, which is a natural sleep aid. One company, Night Time Milk in England, even sells milk from cows milked at night when the melatonin is increased. The milk is marketed as a sleep aid, proving that people will buy anything today.

CHAPTER 5

Eloise, with her wounds cleaned and dressed, is back in business and is refreshing glasses and making small talk. Leyner appears wounded from his laboratory mishap and is quietly sitting cross-legged in the corner, sullenly nursing the dregs of his bottle of tequila. I've

DRUGS AND ALCOHOL

never condoned casual drug use, but I almost feel compelled to spike the punch bowl with a strong sedative and

sneak out to find my way home. I resist this evil urge and feel better until I am confronted again by the indefatigable but once fat guy Jeremy Burns.

Jeremy, aside from his Atkins obsession, has never grown out of his penchant for fraternity hijinks. Eloise offers him one of her signature frozen daquiris, but Jeremy only wants a Jell-O shot, a beer bong, or some

Ecstasy. Eloise turns her nose up at his boorish request, and he turns to me to plead for a prescription for some medical-grade marijuana. I explain to him that prescription pot is not available in New York, and that I wouldn't give it to him anyway.

Jeremy is not ready to give up and asks, "Then can you get me any of that shit that Rush Limbaugh takes?"

"OxyContin," I reply.

"Yeah, yeah," he says. "Oh and also some of that stuff that Matthew Perry and Brett Favre do."

"Vicodin," I reply again.

Jeremy says that he already has plenty of Vicodin and asks if maybe I can just get him a little ketamine.

I am becoming exasperated and I realize that I have an opening. "Jeremy, you know that ketamine is a potent horse tranquilizer . . . and that guy over there is a veterinarian," I say, pointing to a portly, balding gentleman in the next room.

Jeremy rushes off, as I breathe a sigh of relief.

With a culture dedicated to the use and abuse of caffeine, nicotine, alcohol, and an endless array of illegal substances, questions abound about the safest and quickest ways in which we can intoxicate ourselves and how to avoid the dreaded hangover.

"BEER BEFORE LIQUOR, NEVER SICKER/ LIQUOR BEFORE BEER, NEVER FEAR"?

This one isn't all that clear. Or maybe it's because of those drinks we just had.

The biggest problem with this rhyme is that nobody seems to remember how it goes. As for the science, there is no research to prove or disprove it.

One theory about this little ditty attempts to explain that the carbonation in beer causes increased alcohol absorption. There is no proof that this is true. Nor should you believe that coffee will help you with a hangover or that bread will absorb the alcohol in your system. Only time will cure your pain as you wait for the alcohol to leave your bloodstream.

Intoxication is defined as a blood alcohol level of 100 mg/dL (.10%). In adults, the level usually falls about 15 to 20 mg/dL per hour. Everyone metabolizes differently, but on average it would take about six to eight hours for you to return to normal from a mild drunken state.

Blood Alcohol Concentration	Symptoms
.02%	light-headed
.05%	mild euphoria
.08%	loss of critical judgment
.10%	lack of coordination and balance
.15%	disorientation
.20%	vomiting

.30%	drunken stupor
.40%	coma
.45%+	death

Simply put, alcohol causes intoxication, so the more you drink, the sicker you get. It doesn't have anything to do with the order in which you tend to chug your beer or wine.

As for the dreaded hangover that follows, it is caused mainly by dehydration and interrupted sleep. The sleep and water that will ultimately cure you are not as interesting as some of these famous hangover cures:

1. The Prairie Oyster (olive oil, tablespoon of tomato ketchup, one egg yolk, salt and pepper, Tabasco, Worcestershire sauce, vinegar or lemon juice).
2. Cold pizza
3. IV fluids (helps to date an M.D. or paramedic)
4. The hair of the dog that bit you (i.e., the blessed Bloody Mary)
5. Vitamins B and C
6. And the most effective, and most expensive, kidney dialysis

CAN POPPY SEEDS MAKE YOU TEST POSITIVE FOR HEROIN?

If it's the Jewish holiday Purim and you plan on competing in the Olympics, you may want to think twice before gorging on poppy seed hamantaschen. Eating enough poppy seeds can cause your urine to test positive for opiates. It is difficult to say how many poppy seeds you need to eat to fail your drug test, but some reports have stated that three poppy seed bagels, for example, could generate a positive test result. Pastries and cookies that contain heavy amounts of poppy seeds, like hamantaschen, could also lead to a positive test. There is an additional test that looks for certain chemicals present in heroin that are not present in poppy seeds. So, your athletic future really will depend on the exact test you are taking.

What is the poppy seed–heroin connection? Cultivated poppies are the source of opium, from which morphine and heroin are produced.

WHY DO YOU GET THE MUNCHIES WHEN YOU ARE STONED?

Answer: To keep Domino's and Frito-Lay in business.

Marijuana is the most commonly used illicit drug in the United States. The main active chemical in marijuana is THC (delta-9-tetrahydrocannabinol), or The High Causer. THC falls in the category of chemicals called cannabinoids.

A study in the April 2001 issue of *Nature* helps us to better understand how marijuana causes users to have an increased appetite, the famous "munchies." Molecules called endocannabinoids, marijuanalike chemicals present in our own brain, bind with receptors in the brain and activate hunger. These endocannabinoids in the hypothalamus of the brain then activate cannabinoid receptors that are responsible for maintaining food intake. The chemicals from marijuana bind to these cannabinoid receptors and cause the munchies. Sound complicated? Maybe you're too stoned to understand. Go eat some cookie dough.

CAN A HAIR SAMPLE BE USED IN A DRUG TEST?

Hold the Rogaine. If you are bald, there may be another advantage besides the cost savings on hair products. You won't have any hair to offer for your drug test.

As drugs are ingested into the body, they circulate in a person's bloodstream. Trace amounts of these drugs or the drug metabolites are deposited in the hair follicle. As the hair grows, they remain stored in the core of the hair shaft.

When a person is tested, samples are taken at various levels in the hair shaft so that a reasonably accurate approximation can be made of how long ago a particular drug was used. Drugs or drug metabolites cannot be washed, bleached, or flushed out of the hair follicle.

The major practical advantage of hair testing compared with urine testing for drugs is that it can show that drug use occurred in the past weeks to months, depending on the length of the hair shaft, versus within only the past two to four days for other tests. Hair analysis is the least invasive of the testing methods but might not reveal recent use. Blood analysis is the most accurate but definitely invasive. Urine analysis is typically the least expensive and can detect infrequent or a recent single use. Urine analysis is the most commonly used form of drug testing.

So, if you constantly hit the bong, you might want to consider shaving your head.

WILL A SHOT OF BOURBON CURE A COLD?

There has been much discussion and research over the years on the health benefits of alcohol. In the 1920s the "Guinness Is Good for You" campaign in the United Kingdom made people believe that this famous Irish stout had health properties. The slogan stemmed from intense scientific market research: people told the company that they felt good after a pint, and the slogan was born.

Echinacea, vitamin C, zinc, and chicken soup, as well as a stiff belt, have all been postulated to prevent or cure the common cold. Unfortunately, there is no strong evidence for any of these choices. There are many other home remedies, several of which include brandy or whiskey. A friend has her own recipe, combining vodka and orange juice into a screwdriver, as her own special cold cure. Most likely the buzz just helps you forget how bad you feel.

DOES PUTTING SOMEONE IN A SHOWER OR GIVING HIM OR HER COFFEE STOP A DRUG OVERDOSE?

Alcohol is the most common drug that leads people to throw their friends in the shower or force-feed them coffee. Time is the only thing that will sober up a drunk person. Coffee, showers, exercise, sweating it out, fresh air, or any other method will not increase the rate at which alcohol is eliminated from the body. The liver just needs the time to metabolize the alcohol.

As for other, more hardcore drugs, the coffee won't help, but keeping someone awake until help arrives could be a lifesaver. Heroin and other opiates cause you to stop breathing, and this leads to cardiac arrest. Remember, coffee is only a temporary measure and medical help should be sought for any drug overdose. If someone stops breathing, you should begin CPR. The shower is probably unnecessary, and a big waste of precious time.

WHY DO YOU THROW UP WHEN YOU DRINK TOO MUCH?

Vomiting from excessive drinking is simply your body's way of getting rid of the toxins in alcohol quickly. Vomiting is not a bad thing in this case, but repeated hurling can lead to potentially life-threatening dehydration and electrolyte imbalances. There is also the danger of choking on vomit, like the guy from Led Zeppelin.

The urge to vomit comes from two anatomically and functionally separate units—a vomiting center and a chemoreceptor trigger zone. The vomiting center, which has overall control of vomiting, is located in the part of the brain called the medulla. The chemoreceptor trigger zone, which sends signals to the vomiting center, is found in the fourth ventricle of the brain. The ventricles are a system of four communicating cavities in the brain that are filled with cerebrospinal fluid. Alcohol probably acts on the chemoreceptor trigger zone.

In the hospital vomiting is referred to as emesis, but many doctors prefer these more colorful terms:

puke

barf

uneat

blow chow

ride the porcelain bus

pray to the porcelain god

technicolor yawn

toss your cookies

lose your lunch

feed the fish

spill the groceries

DOES TAKING ECSTASY CAUSE YOU TO LOSE YOUR MEMORY?

This is a question that gets asked a lot, both because of the increasing popularity of Ecstasy and the fact that people keep forgetting that they asked in the first place. So, yes, Ecstasy probably does cause memory loss.

Ecstasy, or MDMA (3–4 methylenedioxymethamphetamine), is a synthetic psychoactive drug chemically similar to the stimulant methamphetamine and the hallucinogen mescaline. Some refer to it as a "designer amphetamine."

One of the major results from the use of Ecstasy is that both in short-term and long-term use it could have serious effects on brain cells. Specifically, Ecstasy harms neurons that release serotonin, a brain chemical thought to play an important role in regulating memory and other important functions. Case reports and interviews with Ecstasy users report memory loss, depression, alterations in sleep, and anxiety. Memory deficits seem to persist even after stopping the use of Ecstasy.

DOES DRINKING KILL BRAIN CELLS?

In the process of researching this book, the authors had a working dinner, and in the course of our hard work, we consumed large quantities of beer, wine, and tequila. As we stumbled down the street, Mark insisted that he was fine to take the train home. Better judgment prevailed, and I wrestled him into a cab. The cab drove away and I started to walk home. A block later, I came upon the cab and found Mark fumbling with his empty wallet in the backseat. I opened the door. Mark, with no memory of our evening, said, "What are you doing here?" Alcohol surely kills brain cells.

To properly answer this question we must separate light to moderate drinking from heavy drinking. We also need to separate temporary from permanent damage.

In general, alcohol doesn't specifically kill brain cells but alcohol damages the dendrites, the small branches that extend from the cells and receive information. The mechanism of action for intoxication is multifactorial but the end result is the slurred speech, clumsiness, slow reflexes, and loss of inhibition that we associate with being drunk. This damage isn't permanent in light to moderate usage. This means you can have one to seven drinks a week and be just fine.

Heavy alcohol use does clearly cause neurological damage. CT scans of chronic alcoholics can show brain atrophy and studies have shown that heavy use damages retrospective memory. Alcoholism can produce Wernicke-Korsakoff syndrome, which is caused by deficiency of the B-vitamin thiamine; alcohol decreases the absorption of this vitamin, and alcoholics also don't have the healthiest diets. Patients with this condition have symptoms such as confusion, delirium, disorientation, inattention, memory loss, and drowsiness. If thiamine is not given promptly, the syndrome may progress to stupor, coma, and death.

WHY DOES THE BED SPIN AFTER A LONG NIGHT AT THE BAR?

Nothing is worse than the moment when you hit the sheets and the room starts to spin. Trying to explain why this happens causes almost as much dizziness.

The vestibular system is a complicated network of passageways and chambers within the inner ear, all of which work together to control equilibrium and balance. Inside there are tubes and sacs that contain different fluids, each of which has a different composition. When you are healthy, and both sides of your vestibular system are functioning properly, both sides send symmetrical impulses to the brain. When someone gets very intoxicated, the alcohol changes the density of the blood and this affects the intricate system of balance. That is when the spinning starts. This is very similar to the condition called vertigo.

WHY DO YOU SNORE SO LOUD WHEN YOU ARE DRUNK?

There is a common serenade in any emergency room. The coarse snore of the regular alcoholic fills the air. Normally, we just ignore it. But sometimes, too much alcohol actually impedes the breathing process. We fix this easily with a short small rubber tube in the nose, an aptly named nasal trumpet. Alcohol increases snoring by relaxing the muscles that hold the throat open, allowing the soft palate tissue and uvula to flutter more as air passes.

IS RED WINE REALLY GOOD FOR YOUR HEALTH?

Finally, some good news.

Historically there has been a belief that wine has medicinal properties. Hippocrates and Thomas Jefferson both considered wine an important part of their health regimens. Louis Pasteur, the famous French biologist, said, "Wine is the most healthful and hygienic of beverages."

There now is an enormous amount of research about what has been called "the French Paradox," that despite a diet rich in fats there is a lower-than-expected prevalence of cardiovascular disease among the Gauls.

Scientific studies have linked this surprising fact to the moderate consumption of alcohol, specifically red wine. Red wine has also been linked to a reduced risk of some cancers, atherosclerosis, heart disease, and even the common cold.

So drink a whole bottle tonight. Your bed will spin but you probably won't have a heart attack.

DOES MARIJUANA HELP GLAUCOMA?

There are some important medical uses for marijuana, and some of these lead to solid arguments for legalization. However, the use of marijuana for glaucoma does not appear to have any benefit over available medications.

Marijuana does reduce pressure in the eye, but in order to sustain this reduction you would have to smoke about ten to twelve joints a day. Your eye pressure might be lower but you will be too stoned to get anything else accomplished except naked guitar playing, gluttonous pork rind consumption, or deriving profound meaning from Rob Schneider films.

SHOULD YOU DRINK BRANDY WHEN YOU HAVE FROSTBITE?

The arrival of the Saint Bernard with the little cask around its neck is a heartwarming image but drinking alcohol to warm yourself or prevent frostbite doesn't make any medical sense, we're sorry to report. Alcohol consumption actually can be dangerous in these conditions as it decreases blood circulation and thus can enhance heat loss and impair shivering.

CAN YOU GET HIGH FROM LICKING A TOAD?

Poor, sad toads. They always seem to take a backseat to the frogs. Frogs get kissed and turn into princes, and toads just get to cause warts. Well, here is some good news for toads.

Toads do not cause warts. Toads do, however, produce a protective substance in the parotid gland behind the eyes. This toxin can make animals, such as dogs, very sick and can be irritating to the human eye. But some people go way beyond touching toads and actually lick them in an attempt to get high from a "psychedelic" substance supposedly found on its skin.

The species known as the Bufo toad does have a psychedelic substance on its skin. This substance is similar to serotonin and LSD and can cause hallucinations. Be careful when trying this method because some people have been arrested for toad licking.

WHY DO PEOPLE SEEM MORE ATTRACTIVE TO YOU WHEN YOU ARE DRUNK?

Straight from the Department of Psychology at the University of Glasgow, a paper entitled "Alcohol Consumption Increases Attractiveness Ratings of Opposite-Sex Faces: A Possible Third Route to Risky Sex," thus proving that beer goggles do exist. Feel free to use this paper to excuse your bad behavior.

CHAPTER 6

Leyner seems to have rebounded from his brief period of

remorse and sorrow, and is now back to his crazed

ways. Tequila in hand, he is delivering a rambling

quasi-coherent lecture about cultural differences in

post-defecation hygiene. The audience is appalled,

BATH-ROOM HUMOR

yet raptly entranced by his

scholarly scatological solil-

oquy. As Leyner continues,

a hand pops up in the back

of the room. The hand

belongs to Joel Blake, a celebrity orthodontist, who

starts to ask a question but begins to stammer as tears

well up in his eyes.

Leyner moves through the crowd with the style

and empathic grace of Oprah Winfrey, grabs his hand,

and says, "It's okay Joel, you can tell us, you are among

friends."

"I wipe standing up!" Joel blurts out.

There is a cackle from the gallery but Leyner silences the offender with an icy stare.

"We need to honor everyone's Way of Wiping," Leyner says serenely, as he hugs Joel.

The bathroom and all that occurs behind closed doors may be the final taboo. Yet when placed in a comforting environment or in a locker room, people will share their secrets often to unfortunate results.

CAN YOU DRINK YOUR OWN URINE?

Thanks to our wonderful democratic society, you can do whatever you want. The better question is, Why would you want to drink your own piss?

Drinking small amounts of your own urine is probably safe. It is made up of 95 percent water, 2.5 percent urea, and 2.5 percent salt, other minerals, hormones, and enzymes. Actually, some folks consider it to have therapeutic properties. Ask the Chinese Association of Urine Therapy. They say urine is sterile, antiseptic, and nontoxic.

For serious yoga practitioners, drinking one's urine is called amaroli. One of the most famous users of urine therapy was the prime minister of India from 1977 to 1979, Morarji Desai. At the celebration of his ninety-ninth birthday, Desai attributed his longevity to drinking urine on a daily basis. But, we plan on sticking to morning coffee, a good glass of cabernet, and an occasional Yoo-hoo, even if it knocks a year or two off our life spans.

WHY CAN YOU IGNITE A FART?

- An average fart is composed of about 59 percent nitrogen, 21 percent hydrogen, 9 percent carbon dioxide, 7 percent methane, and 4 percent oxygen. Less than 1 percent of its makeup is what makes a fart stink.

- The temperature of a fart at its time of creation is 98.6 degrees Fahrenheit.

- Farts have been clocked at a speed of ten feet per second.

- A person produces about half a liter of farts a day.

- Women fart as much as men.

- The gas that makes your farts stink is the hydrogen sulfide gas. This gas contains sulfur, which is the smelly component. The more sulfur-rich your diet, the more your farts will stink. Some foods that cause really smelly farts include beans, cabbage, cheese, and eggs. Also soda.

- Most people pass gas about fourteen times a day.

All are important facts, but back to the question: Is it really possible to ignite farts?

The answer to that is yes!

The flammable character of farts is due to hydrogen and methane. The proportions of these gases depend largely on the bacteria that live in the human colon that digest, or ferment, food that has not been absorbed by the gastrointestinal tract before reaching the colon.

There is some danger associated with igniting flatulence. Fraternity guys don't seem to care.

WHY DO YOU GET ALL "PRUNEY" AFTER A LONG BATH?

There is nothing like a long soak in a bath to relax your soul. The problem is that you have to deal with the ghastly sight of your hands and feet after exiting. The simple answer for why this occurs is that our outer layer of skin (the epidermis) absorbs a little bit of water when we soak too long in the tub. Voilà! Old lady flesh!

The skin on the feet and hands is thicker than the skin on the rest of the body and therefore makes any changes more noticeable. As the epidermis expands, the layer below it, the dermis, does not swell, so the epidermis buckles in areas. Lovely, right?

IS IT MORE SANITARY TO BE SPIT ON OR PEED ON?

There is no specific course in medical school to deal with all the secretions that you find yourself faced with as a doctor. It is definitely a rude awakening to find yourself being coughed on, spit on, and even urinated on. All doctors have been doused in a variety of bodily fluids.

One wonderful evening in the ER, I heard a nurse screaming. I found her desperately trying to keep a drunk patient who had passed out from hitting the floor. He was not a small man, and the dead weight was too much to manage. The only way I could get him back on the stretcher was to grab him from behind and throw myself on the stretcher with the patient on top of me. Simple. I could then just roll him over. I unfortunately didn't plan on him using me as a bedpan the instant we hit the bed.

This is disgusting, of course, but when faced with the option of being urinated on or spat on, I would choose urine. No, this is not a fetish. Normal urine is sterile. It contains fluids, salts, and waste products, but it is free of bacteria, viruses, and fungi. It is not always fragrant, but is certainly cleaner than spit. Spit contains large amounts of bacteria and thus is filthy.

WHY DO BEANS GIVE YOU GAS?

It is unbelievable how much information there is available about farts. Flatulence is the subject of numerous medical studies, books, and CDs. One company even makes a fart filter and underpants designed to absorb odor. But among all this gaseous information it always comes back to the bean, the most famous farting food.

Beans contain high percentages of sugars (oligosaccharides) that our bodies are unable to digest. When these sugars make it to our intestines, bacteria go to work and start producing large amounts of gas. We also form gas from other sources, including the air we swallow, gas that seeps into our intestines from the bloodstream, and carbon dioxide formed from saliva reacting with stomach acid.

There is some help available for those who can't handle their beans. A product called Beano is readily available. Beano contains a food enzyme extracted from mold, one alpha-galactosidase, that helps to break down the complex sugars in gassy foods. Another method is to soak beans before you cook them, as this cuts down on their gas-producing power if you then discard the water. Unfortunately, you also lose some water-soluble vitamins by doing this.

Other flatugenic foods are broccoli, brussels sprouts, cooked cabbage, raw apples, radishes, onions, cucumbers, melons, coffee, peanuts, eggs, oranges, tomatoes, strawberries, milk, and raisins.

Notice the abundance of vegetables on the fart-producing list. That is why those vegetarians are always passing wind in yoga class as they contort themselves into weird positions.

WOULD YOU DIE IF YOU ATE YOUR OWN FECES?

There is a psychiatric illness called coprophagia, the eating of one's own feces. It is an uncommonly reported symptom that can be seen in patients with schizophrenia, alcoholism, dementia, depression, Kluver-Bucy syndrome (ask Mark), and obsessive-compulsive disorder. Scatolia, the smearing of feces, is often seen in psychiatric hospitals. High-functioning individuals may sometimes exhibit coprophagia as part of a paraphilia or abnormal sexual arousal disorder. There are even some claims that Eva Braun urinated and defecated on Adolf Hitler. Sexy!

You can get very sick by eating feces. It shouldn't be fatal, but complications from snacking on shit include hepatitis, oral infection, abscess, and a variety of other infectious diseases. Besides that, think of the morning breath.

....

12:05 P.M.

Leyner: Be right there . . .

Gberg: OK, I gotta run to the bathroom.

12:15 P.M.

Gberg: I have returned.

Leyner: Did you wipe standing up?

Leyner: Some people do, I've heard . . .

Gberg: Stop mocking me. You know I am sensitive about being a stander.

Leyner: I'm sorry . . . you know sometimes I pee sitting down . . . out of pure laziness.

Gberg: That is what they should teach you in school.

Gberg: Bathroom etiquette.

Leyner: They should teach boys that they don't HAVE to stand up . . . that it's an option.

12:20 P.M.

Leyner: When my niece was a little girl she said a great thing once on the way back from a little skiing excursion in Lenox, Massachusetts.

Gberg: And . . . what were these words of wisdom?

Leyner: It was quiet in the car and all of a sudden she piped up, "I didn't fart . . . but I'd open a window if I were you."

Gberg: With all the new technology, they should make an automatic sensor that senses the gas and opens the window.

Leyner: I hate going into a bathroom in a fancy restaurant where they have an attendant in there.

Leyner: A men's room seems to be the one place on earth where Emersonian self-reliance should be the rule.

Gberg: I know, I don't really need assistance getting the paper towel out of the dispenser.

Leyner: There's really nothing that goes on in a men's room that I can't handle myself.

Gberg: You end up feeling so guilty that you have to give the poor bastard a tip.

12:25 P.M.

Leyner: You know that expression for waiters and cooks—when they spend their day off at the place they work?

Leyner: Bellman's holiday or something?

Gberg: ??

Leyner: Wonder if there's an equivalent for men's room attendants?

Leyner: Look that expression up online, will you . . . bellman's holiday.

Gberg: They probably can't urinate or defecate at home because it reminds them of work.

Leyner: What about the people who check stool samples all day, like at that place Jetti Katz, you know that lab?

Gberg: What the hell are you talking about?

Leyner: There's a lab I went to once when my cousin, my gastroenterologist, thought I might have picked up some exotic parasite in Tierra del Fuego.

Gberg: Jetti Katz sounds like a performer in the Catskills.

12:30 P.M.

Leyner: So he sent me to this lab that specializes in analyzing stool for parasite eggs . . . Jetti Katz or Jeddi Cats or something like that . . . some place in upper Manhattan.

Gberg: The Jeddi Cats sound like a band.

Gberg: I love their music.

Leyner: Anyway . . . they have these women who work there, and what they do all day is handle hot, steaming fresh stool samples. The whole experience is indelible in my brain.

Gberg: Indelible or inedible?

Leyner: First you down some . . . What's that laxative they give you? . . . It has a catchy name.

Gberg: Go-lytely!

Leyner: Works mighty fast.

Leyner: Know what I'm talking about?

Gberg: Nothing light about it.

Leyner: What's it called? Help me out here.

Gberg: Not Go-lytely?

Leyner: NO.

Gberg: Lactulose, sorbitol, milk of magnesia, cascara, Dulcolax . . .

Leyner: Dulcolax . . . I think.

Gberg: Magnesium citrate.

Gberg: Dulcolax, "the Duke."

Leyner: DULCOLAX, yes!!!!!! Anyway, they make you drink that . . . then, a dozen or so people vie for three bathrooms. It's like some debased Japanese game show.

Gberg: We could be huge in Japan.

Leyner: We already are.

Gberg: Did your books sell internationally?

Leyner: That's the strange thing about Japan. You can be famous there, have malls named after you, etc., etc., and NEVER know it.

Gberg: I am pretty sure there is no Billy Goldberg mall in Kyoto.

Leyner: One of my books has been published in Japan, and all of them in Great Britain, Italy, and France . . . and Chechnya, I think. I'm like the Dr. Seuss of Chechnya.

WILL YOU GET HEMORRHOIDS FROM SITTING ON THE TOILET TOO LONG?

We have no pretensions about this book, and we expect it to be found in that precious spot right next to the toilet. For that reason, we fear we need to warn you that sitting too long on the throne may cause hemorrhoids. Unfortunately, this one's not an old wives' tale.

Hemorrhoids, or piles, are abnormally swollen veins in the rectum and anus. They are similar to the varicose veins you might see on a person's legs at the beach. It's estimated that about one hundred million Americans are currently suffering from hemorrhoids. More than half of the U.S. population develops hemorrhoids by age fifty. The most frequent causes of hemorrhoids are constant sitting, straining with bowel movements (from constipation or hard stools), prolonged sitting on the toilet, severe coughing, giving birth, and heavy lifting. It has also been suggested that the Western diet, which is rich in processed food and lacking in fiber, contributes to hemorrhoids.

Sitting on the toilet too long is problematic because this is the only time that the anus truly relaxes, allowing the veins down there to fill completely with blood. To prevent this problem, you should move your bowels as soon as possible after the urge occurs. If you cannot go right away, pick up our book (we expect it to be toilet reading) but read as you walk. You can always return to finish the job.

WHY DOES POO STINK IF THE FOOD DOESN'T?

We don't want to create any cultural stereotypes here, but most of the bathroom questions came from folks from Down Under. Yes, two Aussie friends seem to ask many questions about their bowels.

Everything that happens in the intestine seems to have something to do with the production of gases and sulfur compounds. The bacteria inside feces is what makes it smell so bad. Specifically, the bacteria produce various compounds and gases that lead to the wonderful smell of a bus station bathroom. The smell of your stool can be affected by medical conditions or your diet. Fatty stools and bloody stools are known to be particularly malodorous. In the hospital, a large, ripe poo is known as a code brown. How's that for real insider knowledge?

WHY DOES POO FLOAT?

Some people seem to be obsessed with the creation of the perfect poo. My brother even called me in to examine his works of art, a true bonding moment for young boys. Another friend described his perfect moment for us when he produced the cobra—one that coiled around and poked its head out of the bowl. There is something about "dropping the kids off at the pool" that makes us all smile. So, laugh if you must, but we're sure you've wondered why some poos are floaters.

It is gas that makes poo float. Increased levels of air and gas make it less dense and therefore cause it to float.

WHY IS POO BROWN?

It is very common to have people ask about the color of their stool to figure out how it relates to disease. There are definitely some color changes that can be cause for concern, but in general assessing stool color is no exact science.

Feces are mostly shades of brown or yellow because of the presence of an orange-yellow substance called bilirubin. Bilirubin combines with iron in the intestine to give the combo a beautiful brown color.

Poo does, however, have a rainbow of possibilities:

- **Black:** A black stool (melena) can mean that blood is coming from the upper part of the gastrointestinal tract, the esophagus, stomach, or first part of the small intestine. Other things that can cause black stool are black licorice, lead, iron pills, Pepto-Bismol, or blueberries.

- **Green:** Green, leafy vegetables contain chlorophyll, which can color the stool green. Green feces can also occur with diarrhea if bile passes through the intestine unchanged. In breast-fed babies, green stool is a normal occurrence, especially right after delivery.

- **Red:** Maroon stool or bright red blood in poo usually suggests that the blood is coming from the lower part of the GI tract. Hemorrhoids and diverticulitis are the most common causes of red blood in the stool. Beets and tomatoes can also make stools appear reddish.

- **Gray:** Diseases of the liver, pancreas, and gallbladder can cause pale or gray stool.

- **Yellow:** One condition that can cause yellow stool is a parasitic infection known as giardia. Giardia also causes significant diarrhea. Another cause of yellow poo is a condition known as Gilbert's syndrome. This is a fairly common genetic disorder that causes an increase in your level of bilirubin. Gilbert's syndrome is rarely dangerous.

WHERE DOES GAS GO WHEN YOU CAN'T FART?

Some people like to think of their lower gastrointestinal tract as a one-way street. One time during a rectal examination during a trauma, a frightened young man screamed out as the doctor was placing his finger in the man's rectum, "Whoa, that's an exit!"

Flatulence follows that same rule. Gas goes out or it simply goes away.

IF YOU ARE STRANDED ON A DESERT ISLAND, SHOULD YOU DRINK SEAWATER OR YOUR OWN URINE?

Seawater is more than three times as concentrated as blood. Humans shouldn't drink salt water because it forces your body to deal with a solution that is more concentrated than its own fluids. In order to get rid of the excess salt, your body must excrete it through the kidneys as urine. The kidneys can only make urine that is less salty than salt water, so if you drink seawater, you'll be peeing a lot and losing an excess of water. This would cause your body to become dehydrated, leaving an excess of sodium in your bloodstream. Water would then leave all your other cells to enter the bloodstream. This would cause the cells to shrink and malfunction. As a result, muscles would become weak and ache, the heart would beat irregularly, you would become confused, and ultimately you would die.

Drinking urine is probably safer than seawater, but the catch-22 is that if you don't have any water to drink, you will become dehydrated and not produce any urine. The best bet is to not get shipwrecked and if you do, hope for rain.

CAN YOU CATCH DISEASES FROM A TOILET SEAT?

By doing research we found reports of gonorrhea, toilet-seat dermatitis (infragluteal eczema), ascaris lumbricoides (roundworm), and enterobius vermicularis (pinworm). We know what you are thinking. After carefully washing our hands, we went back to our computers and came across more information.

Yes, occasionally you can catch something from a public toilet seat but this isn't all that common. Work, on the other hand, may be worse for your health than toilet seats. A microbiologist at the University of Arizona, Charles Gerba, found that the typical office desk harbors around four hundred times more disease-causing bacteria than the average toilet seat. Here is the bacteria count:

telephone: 3,894 germs per square centimeter

keyboard: 511 germs per square centimeter

computer mouse: 260 germs per square centimeter

toilet seat: priceless

WHY DO I HAVE TO GO TO THE BATHROOM IMMEDIATELY AFTER A CUP OF COFFEE?

In our house, we call it the coffee alarm. Nothing is more reliable.

Coffee is definitely known to have a laxative effect. The caffeine in coffee speeds up every system in the body including the bowels. But when used excessively, caffeine can interfere with the bowels' normal contractions and lead to constipation. Decaffeinated coffee does away with the caffeine but it still acts as a bowel irritant.

WHY DO CIGARETTES HAVE A LAXATIVE EFFECT?

There is nothing better, for some, than the morning cup of joe and a cigarette, followed by the "morning constitutional." Caffeinated drinks and nicotine have a laxative effect probably because they stimulate nerves that increase intestinal contraction, so if you had a block of cheese for dinner, this is a wonderful remedy.

If you want to have a cigarette and coffee for breakfast, make sure that you have a clean toilet nearby.

WHY DO YOU HAVE TO PEE WHEN YOU HEAR WATER DRIPPING?

Sorry, pal, sometimes there just aren't medical explanations for things. Nobody in medical school explained why you get the urge to urinate when you pour gas into the generator.

IF YOU STICK A SLEEPING PERSON'S HAND IN WARM WATER, WILL HE OR SHE WET THE BED?

Going to bed at camp always felt like a risky time. The fear of having someone dip your hand in warm water and waking in a puddle was terrifying. There is no clear medical proof to this camp myth but there may be some science behind it. It is known that when someone has trouble urinating, a warm bath can sometimes make it easier to go, perhaps because of a reduction of pressure in the urethra with the increased body temperature during a bath.

A study from Egypt called "Role of Warm Water Bath in Inducing Micturition in Postoperative Urinary Retention After Anorectal Operations" described this so-called thermosphincter reflex in 1993. We still don't know if the hand dip works the same way, but it sure would be funny to picture an Egyptian in the laboratory sneaking up on sleeping volunteers to try to get them to wet their beds.

IS IT DANGEROUS TO HOLD IT WHEN YOU HAVE TO PEE?

My junior high school biology teacher instilled fear in our young hearts when he told us that if we got into a car accident with a full bladder, the bladder could rupture. He was right. In general, a full bladder ruptures more easily than an empty bladder.

This doesn't mean that your bladder will explode if you hold in your urine because your dad, husband, or brother won't make a pit stop.

Our bodies have a nonvoluntary reflex mechanism to prevent our bladder from getting too distended, called the micturition reflex. When our bladder gets distended, there are stretch receptors in the bladder wall that let us know that it is time to go. As we all know, this is not the most comfortable sensation (if you wait too long). These sensory neurons cause contractions that can become strong enough to overcome the muscle tone holding the urethra shut and release all that urine.

WHAT CAUSES THE RUMBLING IN YOUR STOMACH WHEN YOU HAVE TO GO?

Borborygmi: **bor·bo·ryg·mi;** noun, plural : rumbling sounds caused by gas moving through the intestines.

Certainly not a word learned in medical school, or that can be easily used in a sentence.

These rumbling sounds are a normal part of digestion. They are generated from muscular activity in the stomach and small intestine as the food, gas, and fluids are mixed together and pushed through the intestine. This squeezing of the muscular walls is called peristalsis. Many people associate these sounds with hunger because they are louder and echo more when the intestine is empty. Yum!

9:07 A.M.
Gberg: Good morning.
Leyner: All the best to you and yours . . . give me one moment, just finishing an e-mail . . .
Gberg: I'm just making a coffee.

9:10 A.M.
Leyner: OK . . .
Gberg: Coffee and pizza for breakfast.

9:15 A.M.
Leyner: I had a Slim Jim and a fermented mare's milk.

Gberg: It's hard to find good fermented mare's milk these days.

Leyner: People should drink the milk of as varied a miscellany of mammals as possible.

Gberg: Did you ever sample human breast milk back in the day?

Leyner: All those diverse antibodies are good for getting an immune system ready for the coming apocalyptic flu pandemic.

Gberg: I live in fear of the avian flu.

Leyner: No . . . never sample that mamma milk . . . I've tried to keep the birthing process and my sex life as far apart as possible.

Leyner: The whole avian flu thing seems like some Hollywood pitch . . . like Hitchcock's *The Birds* and . . .

Gberg: So, my brother, we need to finish this bitch and move on to bigger and better things.

Leyner: What was that epidemic movie?

Gberg: *Outbreak* with Dustin Hoffman.

Leyner: What are doing right now . . . are we IMing about something we need to be IMing about?

Gberg: Not yet, really just getting loose.

Leyner: Otay.

Leyner: Tomorrow seems not so good to work . . . is Thursday any good?

Gberg: I am doing stretches at the same time.

Leyner: Tell me about Thursday.

Gberg: Can't do Thursday. Working.

9:30 A.M.

Leyner: One of those six-pack ab cover
mags . . . that men should do the recumbent
bike in the gym and not the regular one
that puts pressure on the cajones and the
tender perineum.

Gberg: There is something about bike rid-
ing that can damage the pudendal nerve and
affect your front end lifter.

Gberg: The tender perineum—who wrote that?
Fitzgerald?   •

Leyner: I love when you conflate urology
and heavy machinery.

Leyner: "The Tender Perineum" . . . yes,
yes . . . the unfinished F. Scott masterpiece . . .

Gberg: I never knew I could conflate.

Leyner: Poor slob never had the chance to
work out the denouement . . .

Leyner: I heard some woman talking about
morning erections the other day . . .

Gberg: Did you just spontaneously spell
denouement correctly?

Leyner: At a supermarket in L.A.

Gberg: I was wondering where you heard
that?

Leyner: Yes, I spontaneously spelled it correctly . . . it's the coffee.

Gberg: I didn't know that people actually spoke in L.A.

Leyner: This woman . . . enormous plastic L.A. tits and the face of a wizened gargoyle . . . said she won't touch a morning erection . . . because it's not "for me" (she said) . . . it's just a "reflex."

9:35 A.M.

Leyner: I guess people want to feel they've "earned" a change in some other human's physiognomy.

Gberg: L.A. is such a bizarre place. New Yorkers would take advantage of any erection. Why waste a good thing?

Leyner: You know those commercials for that new Viagra . . . whatever it's called?

Gberg: Cialis. Damn, those people seem relaxed and happy.

Gberg: We should start our own pharmaceutical company.

Leyner: Why do they say—at the end of that ad—that you should report erections that last over four hours to your doctor?

Leyner: Maybe you should report them to the police?

Leyner: What's the danger of a four-hour erection anyway?

Gberg: Priapism, my friend, priapism.
Gberg: Very painful and can cause perma-
nent damage to the penis.
Leyner: Can you get a permanent erection?
Gberg: Me, personally?

9:40 A.M.
Leyner: That's funny.
Leyner: Porn stars are said to be able to
get their erections back quickly . . . it's
a vocational skill in high demand in the
industry . . .
Leyner: What accounts for the difference
in the refractory time for various men?
Leyner: Is that what that's called?
Gberg: It sounds so scientific.
Leyner: That's the right term! I just
looked it up. I'm so smart . . . Don't you
think?
Gberg: You could write a scientific
article, "The Refractory Erectile Period
in the Porn Industry."
Gberg: Just as important as making it go
up is making it go down. For those embarrassing public moments.
Leyner: Speaking of porn.
Leyner: Did we ask this in the book: Can
women ejaculate?
Gberg: All we are doing is speaking of
poop, porn, and penises.

Gberg: So sophomoric.

Gberg: Yes, and they can.

9:45 A.M.

Leyner: . . . that we're hard-wired to launch our genes into the future before we decay in a puddle of excrement and putrescence???

Leyner: You think THAT's sophomoric?

Leyner: That's the whole comic tragedy of life!

Leyner: And the central thesis of our book, yo.

Gberg: What is the thesis of our book?

Leyner: The intertwining cosmic threads of poop and porn.

Leyner: That's string theory, ever hear of it?

Gberg: I am very slow on the keyboard this a.m.

Gberg: I think my head is going to explode.

Leyner: That we valiantly attempt to create poetry and architecture and pass along culture and bequeath our genetic heritage, ALL in the face of certain decrepitude and the abject indignities of old age and DEATH.

Leyner: It's a grim struggle each and every day to maintain my dinginity in the face of "reading glasses."

Leyner: Dignity.

Leyner: I misspelled in my passion.

Gberg: Hold on 1 sec.

9:55 A.M.

You left the chat by logging out or being disconnected.

....

CHAPTER 7

The flurry of bathroom talk was cathartic, leaving everyone feeling purged and invigorated. Joel, with a newly found confidence, is leading a small group in a game of charades. Leyner, with the nearly empty tequila bottle in one hand and a fat Cohiba in the other, is gesturing madly and at-

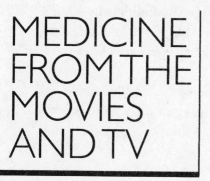

MEDICINE FROM THE MOVIES AND TV

tempting to act out a scene from *Gone With the Wind*. I mistake Leyner's gesticulations for a focal seizure and I run across the room to administer first aid. The group assumes that this is all part of the clue giving and continues to shout out movie titles. Leyner's face is contorted in a bizarre grimace as I assist him to the ground and protect his airway with a head tilt and jaw thrust. Leyner is now scowling and I realize this is not a seizure as Cinderella incorrectly guesses *Spartacus*.

Joel shouts out, *"Vision Quest!"*

And Jeremy quickly replies, "Dude, they're not wrestling, I think they are in love."

Joel quickly responds, *"The Birdcage!"*

And Cinderella guesses, *"Willy Wonka and the Chocolate Factory."*

Jeremy turns to me and blurts out, "Are Oompa Loompas orange from eating too many carrots or are those little bastards just using too much self-tanner?"

Before I can respond, Eloise saunters over and—astonishingly—says in a slow, wistful drawl, "It's got to be *Gone With the Wind*. I'd have a fit too if my ten-million-dollar Charleston bungalow burned to a crisp. Oh poor beautiful Tara."

People often leave the movie theater filled with questions about what they've seen on the screen. After a thrilling episode of *ER*, I can always expect to get a call.

IS THE SHOW *ER* ACCURATE?

Accurate: **ac·cu·rate;** adjective.

1. Correct in all details.
2. Free of mistakes or errors.

Of course it's not completely accurate! It's TV.

But the writing staff does capture the general controlled chaos of an ER. They deal with real medical cases but their medical depictions are always embellished to add a little extra Hollywood flair.

I did my residency in emergency medicine in Los Angeles, when *ER* was just starting and the writers often came by our hospital looking for new ideas. One patient I saw there was portrayed in an early episode, and highlights the writers' taste for the dramatic. One day, a baby was playing with a coat hanger and the tip of the hanger got stuck in the back of his throat. The paramedics carefully brought the baby in to our ER with the hanger dangling from his mouth. This was, of course, very dramatic looking, and we all rushed over immediately. The child was scared but was breathing fine and my fellow doctors and I did our best to just leave him alone and keep him calm. (Rule #1 of medicine: do no harm.) An X ray showed that the tip of the hanger was superficially caught up on the back of the child's throat. Now . . . for the big dramatic ending: we simply reached inside and removed it. Case over!

On *ER*, however, when the glamorous doctors tried to remove the hanger from the child's throat, the baby started to bleed profusely. After an emergency tracheotomy, some miraculous bedside surgery, and a little on-screen romance, this child was just barely saved.

DOES THE KIND OF AMNESIA YOU SEE IN THE MOVIES REALLY EXIST?

Amnesia seems to be one of the favored topics of screenwriters for an easy plot twist. The list of "amnesia movies" can go on and on: *The Bourne Identity, 50 First Dates, Desperately Seeking Susan, Eternal Sunshine of the Spotless Mind, Overboard, Spellbound, Total Recall,* to name just a few.

The amnesia that doctors see is very different than the big-screen version.

Amnesia is best defined as a failure to retrieve information or to place information in an appropriate context. Amnesia rarely erases memory of all past events. No one suffering from amnesia actually goes through the rest of his or her life without an identity or any knowledge of the past. Memory loss is usually temporary and only involves a short time span of the person's life.

TYPES OF AMNESIA

- **anterograde amnesia:** Inability to remember ongoing events after the incidence of trauma or the onset of the disease that caused the amnesia. Anterograde amnesia often occurs following an acute event such as a trauma, a heart attack, oxygen deprivation, or an epileptic attack.

- **retrograde amnesia:** Inability to remember events that occurred before the incidence of trauma or the onset of the disease that caused the amnesia. Retrograde amnesia is often associated with neurodegenerative pathologies such as senile dementia and Alzheimer's disease.

- **emotional/hysterical amnesia (fugue amnesia):** Memory loss caused by psychological trauma such as a car crash or sexual abuse. Usually it's a temporary condition.

- **lacunar amnesia:** Inability to remember a specific event.

- **Korsakoff syndrome:** Memory loss caused by chronic alcoholism.

- **posthypnotic amnesia:** Memory loss sustained from hypnosis. Can include inability to recall events that occurred during hypnosis or information stored in long-term memory.

- **transient global amnesia:** Spontaneous memory loss that can last from minutes to several hours and is usually seen in middle-aged to elderly people.

Here are some movies that Hollywood executives would like to forget: *Gigli, Ishtar, Howard the Duck, The Postman,* and *The Adventures of Pluto Nash.*

WHAT WOULD HAPPEN IF YOU STUCK SOMEONE IN THE HEART WITH A NEEDLE AS IN *PULP FICTION*?

*"No, you don't gotta fucking stab her three times!
You gotta stab her once, but it's gotta be hard enough to
break through her breastbone into her heart, and then
once you do that, you press down on the plunger."*
—Lance (Eric Stoltz), *Pulp Fiction*

No, intracardiac injection is not recommended for routine use during CPR. Translation: ER docs don't ever inject anything directly into people's hearts. There is an uncommon procedure called a pericardiocentesis when a needle is inserted under the breastbone and into the sac around the heart to remove excess fluid. This is done when fluid or blood surrounding the heart is restricting its function. This is only done in serious emergencies.

In *Pulp Fiction*, John Travolta and Samuel L. Jackson are trying to save Uma Thurman from a drug overdose by injecting her with adrenaline. Instead, they should have given her an injection of a medication called Narcan to reverse the effect of the heroin. Adrenaline (epinephrine) is often given during cardiac arrest but only through a vein. Sorry, not as dramatic but definitely less painful.

WHY DOES EVERYONE IN THE MOVIES GO INTO SHOCK?

In the movies and in the dictionary, "shock" can mean to strike with great surprise and emotional disturbance. In medicine, "shock" is a major medical emergency. When doctors talk about shock we are referring to the failure of the circulatory system to maintain adequate blood flow. This requires rapid treatment, otherwise it can lead to death.

There are a number of different causes of medical shock, including bleeding (hypovolemic shock), inability of the heart to pump enough blood (cardiogenic shock), severe infection (septic shock), and life-threatening allergic reactions (ana-phylactic shock). People suffering from shock have low blood pressure, difficulty breathing, a weak rapid pulse, cold and clammy skin, decreased urination, and confusion.

So, the shock that we so often see on-screen should be more appropriately called freaked out.

CAN PEOPLE REALLY WAKE UP AFTER BEING IN A COMA FOR YEARS?

If real life were a soap opera, then the answer to this question would always be yes. Unfortunately, coma is a very serious problem and although people do wake up, the longer they remain in this state, the less likely they are to return to consciousness. This is a very delicate question because doctors can't really predict which patients will wake up and which will not.

If you look carefully in the original movie *Coma,* you will see Tom Selleck, that's right, Magnum, P.I., in a state of suspended animation. Recent movies like *While You Were Sleeping*, *Kill Bill*, and *Talk to Her* also used coma in their stories. The medical story of coma isn't as glamorous as Hollywood portrays.

To begin, there are several different categories of coma, or disorder of consciousness. Consciousness can generally be divided into two main components, arousal and awareness. Coma is defined as a state of unresponsiveness from which an individual has not yet been aroused. Patients in a coma are neither awake nor aware of their surroundings. On average, coma doesn't usually last very long. After several weeks, most patients either regain some level of consciousness and if not are classified as being in a persistent vegetative state. Persistent vegetative state is characterized by complete lack of awareness of self or one's environment. These patients can appear awake and even have their eyes open but are totally unaware of their surroundings.

Another category of consciousness is the minimally conscious state, an intermediate stage of consciousness, which indicates that a patient is somewhere in between a persistent vegetative state and normal consciousness. These patients can show intermittent signs of awareness.

The "locked-in" syndrome is a rare condition that must be distinguished from disorders of consciousness. It is characterized by complete paralysis of the voluntary muscles in all parts of the body except for those that control eye movement. These patients can think and reason but are unable to speak or move.

DO YOU REALLY NEED TO REMOVE A BULLET RIGHT AWAY LIKE THEY DO IN OLD WESTERNS?

It certainly would be dramatic if we had our gunshot trauma patients take a swig of whiskey and bite on their belt as we removed the bullet with a knife that had been sterilized by heating over a fire. I also would love to ride a white horse to work every day but that doesn't happen either.

In old Westerns, there is an urgency involved with removing a bullet, as if this is the life-saving maneuver. In reality, doctors are not concerned with the presence of the bullet but rather the damage that it does on its way in or out. We often see patients who get an X ray for another reason only to find a bullet from a previous injury.

There are some special situations when we worry about leaving a bullet in a person's body. When bullets or fragments are near large blood vessels, nerves (especially the spinal cord), or in a joint, then they can migrate and cause damage. In these cases, bullets are usually removed.

People have also asked whether or not you can get lead poisoning from bullets that are left in the body. In general, lead fragments in soft tissue become surrounded by fibrous tissue and are therefore essentially inert. If a bullet is in a joint, there can be a problem with lead poisoning. A study in Los Angeles in 2002 looked at more than four hundred patients who had bullets retained in their bodies. They found increased levels of lead in a small percentage of patients. Bullets or shotgun pellets are 50 to 100 percent lead and people are more likely to have problems with lead poisoning if there are multiple bullets or multiple fragments in the body. Sorry to disappoint you spaghetti Western aficionados, but the old whiskey-and-leather routine is just for show.

IS THERE REALLY A MEDICATION THAT ACTS LIKE A TRUTH SERUM?

Action heroes like Arnold Schwarzenegger often find themselves faced with an interrogator who uses a truth serum to get the hero to reveal his secrets. In the movies, our heroes are able to resist these potions and hide the truth. Hiding the truth seems to also prepare action heroes for a successful career in politics.

They seem pure fiction, but truth serums do exist. Barbiturates such as sodium amytal and sodium pentothal were first used as truth serums in the early twentieth century. These drugs inhibit control of the central nervous system and were used by physicians to help patients recover forgotten memories or repressed feelings. They are also used for patients with suspected conversion disorder, a condition in which psychological problems produce physical symptoms.

An "amytal interview" is performed by administering a small amount of this drug intravenously. The drug produces a state of drowsiness, slurred speech, and relaxation. This condition makes patients more susceptible to suggestion, allowing the potential to uncover repressed feelings or memories.

Today these interviews are seldom performed. The "truth serum" will not necessarily make you tell the truth. Patients may lose inhibition but will not lose all self-control. Therefore, they are still able to control their behavior and lie. Studies have shown that during these "amytal inter-views," patients often demonstrate a distorted sense of time, show memory disturbances, and have difficulty distinguishing between reality and fantasy, so the line between fact and fiction becomes even more blurred.

WHAT IS ON THE RAGS THAT VILLAINS USE TO MAKE THEIR VICTIMS PASS OUT?

We've all seen it in the movies. The bad guy grabs someone from behind, places a rag over the victim's nose and mouth, and instantly the person slumps to the floor.

This isn't exactly how anesthesia is administered in a hospital, but many people wonder if this rag trick is possible, and if so, what is the chemical on the rag?

Chloroform and ether are the two possibilities. In the mid-1800s, both of these chemicals were being used as anesthetics. Chloroform is the more common substance discussed in these cinematic knockouts but actually doesn't work as swiftly as portrayed. It usually takes several minutes to induce a state of unconsciousness with chloroform. Chloroform also causes a lot of side effects including nausea, vomiting, and skin irritation.

Ether was discovered in the 1500s and later used as an anesthetic. It was also used to treat respiratory ailments such as asthma. Ether became popular and turned into an early party drug that people used to get high. Unfortunately, ether can't deliver that dramatic takedown either.

Another option for the quick movie knockout involves secretly dissolving a pill in a drink. This is often referred to as "slipping a mickey" or a Mickey Finn. Spiked drinks in this sense contain Chloral hydrate dissolved in alcohol. Chloral hydrate is a sedative that is used in hospitals today, often to sedate children prior to procedures. Other modern movie knockout options are the so-called date rape drugs: GHB, Rohypnol, and ketamine.

CAN YOU DIE FROM CHOKING ON YOUR OWN VOMIT, LIKE THE DRUMMER IN *SPINAL TAP*?

In the movie *Spinal Tap*, Eric "Stumpy Joe" Childs, the second drummer in the eponymous band, died in 1974 from choking on vomit. As the movie reveals, "The official cause of death is he died of choking on vomit. It wasn't his own vomit. He choked on somebody else's vomit."

This event is said to have been inspired by the death of John Bonham, the drummer of Led Zeppelin. In 1980, Bonham was found dead after a drinking binge. He apparently had passed out and choked on his own vomit.

This is a true and potentially dangerous possibility with excessive alcohol intake. Death from alcohol abuse most often occurs from aspiration. Aspiration is defined as breathing the vomit into the lungs, which causes the victim to essentially drown.

DO PEOPLE REALLY HAVE MULTIPLE PERSONALITIES, LIKE IN *SYBIL*?

The 1976 TV movie *Sybil* was based on a book of the same name written by Flora Rheta Schreiber. Both of these accounts were based on a real-life patient and psychiatrist but recently there has been controversy over whether the real Sybil actually had multiple personalities at all. Other movies like *The Three Faces of Eve*, *Psycho*, and *Me, Myself & Irene* have also dealt with multiple personality. Once this disorder was popularized in the movies, the number of diagnosed cases increased dramatically.

Multiple personality disorder is no longer the term used to refer to this disease. It is now referred to as Dissociative Identity Disorder (DID). DID is defined as a disorder where two or more distinct personality states or identities alternately control or take over a person's mind. This disorder is a result of many factors, most commonly severe emotional stress.

CAN YOU GET SCARED TO DEATH?

You can scare the pants off someone or be scared shitless, scared stiff, or scared out of your wits. But is it really possible to be scared to death?

There is significant evidence that psychological and emotional stress can increase the likelihood of a heart attack. So it makes perfect sense that the stress of fear could lead to sudden death.

In *The Hound of the Baskervilles*, a famous Sherlock Holmes story that has been adapted into film several times,

Sir Charles Baskerville dies of a heart attack after being frightened by a ferocious dog. An article from the *British Medical Journal* in 2001 entitled "*The Hound of the Baskervilles Effect: A Natural Experiment on the Influence of Psychological Stress on the Timing of Death*" examines whether this phenomenon is actually true. These researchers wanted to show that people are more likely to die of a heart attack when they suffer extreme emotional stress, so they focused on the death rate on the fourth day of the month. In Japanese and Chinese cultures, the number four is associated with death and is feared and avoided. This is not true in other cultures.

When the death rates between Japanese and Chinese Americans and white Americans were compared on the fourth of the month, the researchers found that death rates on this day peaked in the Japanese and Chinese but not in other groups. So it seems as though you can be scared to death—by the number four at least.

CAN YOU DRINK YOURSELF TO DEATH LIKE NICHOLAS CAGE IN *LEAVING LAS VEGAS*?

Even if you don't end up choking on your own vomit, alcohol is still pretty dangerous. The consumption of even small quantities of some types of alcohol, such as methanol or rubbing alcohol, can be fatal.

With ethanol, the alcohol that is found in vodka and wine, for example, too much can certainly lead you to the pearly gates. People often wonder how much alcohol can be life threatening. In medicine, we use the term LD_{50} to describe the dosage or amount of alcohol that causes death in half of the population. The LD_{50} for alcohol is equal to a blood-alcohol concentration of 0.4 to 0.5 percent. That would be about four to five times the amount required to make you legally drunk.

To give an example of how much drinking this means, a hundred-pound person would need to drink about ten drinks in an hour to threaten his or her life. Our bodies tend to protect us from alcohol-related death by vomiting or passing out. The danger occurs when you puke and faint at the same time. If you are brought to the hospital, we will protect your breathing and wait for the alcohol to move out of your system. Stomach pumping for alcohol abuse is a myth since you do that yourself when you puke. Oh, and occasionally when it is a slow night in the ER, the staff will bet on who can guess your blood-alcohol level, just to pass the time. . . .

DOES HYSTERICAL BLINDNESS REALLY EXIST?

On an episode of *King of the Hill,* Hank accidentally sees his mother in bed with her new boyfriend and suddenly loses his vision. In the movie *Hollywood Ending,* Woody Allen's character has the same problem because he is so nervous about the film he has to direct. So, does this sudden blindness really happen outside of the movies and TV?

The answer is definitely yes. And it is not unusual to see these patients in the ER. Hysterical blindness can occur as a result of a psychological stress (a conversion disorder) or someone can intentionally fake blindness for some secondary gain (malingering)—a prisoner who says he can't see in order to try to avoid going directly to jail. It is not difficult to figure out when patients say they are blind but can actually see. We have a simple test that lets us determine whether the eyes are functioning. Using a rotating striped drum, we test for something called optokinetic nystagmus. As the drum spins, normal eyes will be seen moving back and forth.

If a striped rotating drum is not available, you can always use a picture of J. Lo's rear. Move it back and forth, and any normal eyes will follow.

••

Leyner: So . . .

Gberg: Just sent you the blindness question.

Leyner: Okay . . . just got the e-mail . . . I'm reading it now . . . hold on (to something of your own choosing).

Gberg: Just hold on to what you've got. You've got a lot girl, you've got a lot. Got a lo-ovely feelin'. Hang on, hang on to what you've got. I may have left out a "hang on" in that musical interlude. Singing doesn't really work on IM.

Leyner: I think we need to explain just what that test is, then maybe make some joke about something visual that would be almost impossible not to react to, like some starlet's muff for instance, then . . . maybe a joke about what might cause hysterical deafness.

Gberg: Very good. Will do. Let's focus on the intros.

Leyner: Okay . . . let me read through the intro again. . . .

∙∙∙

WHAT WOULD REALLY HAPPEN IF A JUNIOR MINT FELL INSIDE SOMEONE DURING SURGERY, AS IN THE INFAMOUS *SEINFELD* EPISODE?

We're not sure that we can answer this one with any scientific references, and there probably isn't a hospital that would allow you to study the consequences of leaving movie candy inside a patient during surgery. This is not to say that surgeons don't occasionally leave things behind. Surgical sponges and instruments are the most common items left behind, and believe us, it has happened.

In the *Seinfeld* episode, the patient makes a miraculous recovery and it is implied that the mint may have prevented infection. Although there are some reports about using granulated sugar and honey on wounds, having a Junior Mint inside your body is more likely to cause an infection. So, remember to always ask your surgeon to step out of the operating room if he or she needs a snack.

IS IT DANGEROUS TO EAT ANOTHER HUMAN BEING?

One of Mark Leyner's favorite recent news stories is that of Armin Meiwes, a German computer technician who was convicted of murdering someone for sexual pleasure and then eating him over the next several months. Mr. Meiwes had advertised on the Internet for "well-built young men aged eighteen to thirty to slaughter."

Mr. Meiwes in interviews with court psychiatrists said that his fantasies of cannabilism began as a child from watching horror films. For those film buffs who are looking for a viewing list, these movies all involve cannibalism: *Alive; Eating Raoul; The Silence of the Lambs; Hannibal; The Cook, the Thief, His Wife, and Her Lover;* and *Night of the Living Dead.*

So, is it dangerous to eat another human being? I am sad to report that it really isn't that dangerous. Human flesh holds much nutritional value and will keep you alive if your plane goes down and all you have are your fellow, more unfortunate, passengers. Unless you are eating the brain.

A rare disease called Kuru can occur from eating human brains, which killed about 10 percent of the Fore, a New Guinea tribe of cannibals. The Fore would honor their dead by eating them. The brain was reserved for the female relatives and children. Whole villages were wiped out by this rare neurodegenerative disease.

Kuru manifests with muscle weakness and trouble walking. The Fore would then have trouble talking and could no longer stand, sit, or even hold their heads up. Death ultimately resulted from starvation or an infection that developed when they became so sick. Researchers were very interested in this disease because it is very similar to mad cow disease.

••

12:40 P.M.

Gberg: Time really flies when you are typing away at this IM thing.

Leyner: Are you being sarcastic?

Gberg: No.

Gberg: How was the lamb your mother-in-law made last night?

Leyner: The goat you mean.

Leyner: It was great.

Leyner: I love goat and all things goat.

Gberg: I made a mean beef tenderloin last night.

Leyner: Meat, cheese, milk, etc.

Leyner: How'd you make it?

Gberg: In a red wine sauce, tender and delicious.

Gberg: Did I tell you that I added in your favorite story of that German cannibal?

Leyner: That sounds great.

12:45 P.M.

Leyner: I saw that . . . that's essential and indispensable for this book.

Gberg: What an insane story.

Leyner: It's a lot more common than you think. Families tend to keep cannibalism hushed up . . . I had an uncle . . .

Gberg: I will never come to your family's for Thanksgiving.

Leyner: Never mind.

Leyner: I was looking up satyriasis.

Leyner: Speaking of goats.

Gberg: What is satyriasis?

Leyner: It comes from the word "satyr," meaning part man, part goat (fond of Dionysian revelry).

Leyner: Satyriasis: abnormal sexual craving in the male.

12:50 P.M.

Gberg: Says on the Internet that it is caused by extreme narcissism.

Leyner: Really . . . I'm in the high-risk category then.

Leyner: I can spend all day just staring at a single vein on my left bicep.

Gberg: There are treatment options available, medication or . . . I assume that castration is not an option.

Leyner: I'm not taking some horse suppository, son.

Gberg: Maybe some very tight underwear?

. . . .

HOW MANY TIMES CAN YOU BE SHOT AND STILL SURVIVE?

At the end of Scarface, Tony Montana gets shot many times but doesn't lose his ability to spew obscenities. In the hospital we believe that an innocent person will get killed by a single gunshot but the meanest, guiltiest thug can survive multiple gunshots and simply get up, curse at the doctors, and walk out.

The truth is that it really depends on where the bullet hits you.

IS THERE SUCH A THING AS A WEREWOLF?

It happened in An American Werewolf in London, and who can forget Michael J. Fox as Teen Wolf? Lyncanthropy refers to the delusion that one is a wolf. This can definitely be seen in psychiatric illness, but it may be that in some cases this is not a delusion at all. The werewolf legend may have originated out of two medical conditions.

Porphyria is a rare hereditary blood disease. There are two types of porphyria. In one type, cutaneous porphyria, the symptoms can resemble the characteristics of a werewolf. These patients become extremely sensitive to sunlight, grow excessive amounts of hair, and develop sores, scars, and discolored skin. Porphyria also leads to tightening of the skin around the lips and gums, and can make the incisors stand out (think fangs).

Another disease that may have contributed to the werewolf myth is congenital hypertrichosis universalis, sometimes known as human werewolf syndrome. This is another rare genetic disorder that is characterized by excessive hair growth over the whole body, including the face. If you travel to Austria, you can see portraits of the first family discovered with this condition in Ambras Castle near Innsbruck.

So, there isn't really such a thing as a werewolf, but there is a possible medical explanation of how the stories began. Sorry, we don't have a medical explanation for Dracula, Frankenstein, or the Abominable Snowman, but we'll do some research and include it in our next book, *Why Are Women Smarter?*

CAN YOU REALLY EXPLODE FROM EATING TOO MUCH?

In *Monty Python's The Meaning of Life*, a man eats a massive feast, but one wafer-thin dinner mint puts him over the edge. He explodes all over the restaurant. With the obesity epidemic in our country, we have a great deal to worry about, but don't expect to see people exploding at McDonald's. People won't explode from overeating, but if you eat too many Big Macs, you can rupture your stomach.

Stomach rupture, or gastrorrhexis, is a rare condition, although it has been reported to occur from eating too much. In a 2003 issue of *Legal Medicine,* Japanese scientists Ishikawa et alia, reported the case of a forty-nine-year-old man who was found dead in a public restroom after his stomach exploded from eating too much. There is no mention of what his last supper was, and therefore no reason to suspect Pop Rocks and Coke (see chapter 8, page 192).

DO PEOPLE EVER HAVE WEBBED HANDS AND FEET LIKE THE MAN FROM ATLANTIS?

Does anyone else remember the Man from Atlantis? Patrick Duffy (Bobby Ewing from *Dallas*) played the last man from the legendary underwater city of Atlantis. He had webbed feet and hands and gills instead of lungs. This fantastic show only lasted for one season, but it inspired a TV junkie to ask if people could really have webbed hands and feet.

The answer is yes! People can have webbed hands and feet. Actually, it is more common than you may think, occurring anywhere from one in one thousand to one in two thousand births. There are two types of webbing: syndactyly is when two fingers or toes are fused or webbed; polydactyly involves the webbing of more than two fingers or toes. We all start life with hands and feet that resemble a duck, and between the sixth and eighth week of development, our fingers and toes separate. The failure of this separation is what leaves you looking like the Man from Atlantis.

WHY DO YOU SEE STARS WHEN YOU ARE HIT IN THE HEAD?

It always happened to Wile E. Coyote. The Road Runner drops an anvil on his head and then the poor coyote sees stars circling his head. Not only does this happen in cartoons but it is actually a sign of a concussion. A concussion is simply when an injury to the head causes your brain to move around inside your skull.

As for the stars, what probably happens is that the portion of your brain that is responsible for visual information, the occipital lobe, bangs up against the side of the skull.

WHAT WAS WRONG WITH THE BOY IN *THE BOY IN THE PLASTIC BUBBLE*?

In 1976, one year before John Travolta was dancing his way through *Saturday Night Fever*, he was in *The Boy in the Plastic Bubble*. The film was based on a true story of a boy suffering from a rare inherited disease called Severe Combined Immunodeficiency Disease (SCID). SCID is now often referred to as "bubble boy" disease, thanks to this cinematic tour de force.

Severe Combined Immunodeficiency is a life-threatening syndrome in which there is a defect in the white blood cells that protect us from infection. This lack of a functioning immune system leads to frequent severe infections. Patients are usually diagnosed before they are three months old and if untreated the syndrome can be fatal. New treatments such as stem cell or bone marrow transplantation can save many patients. Gene therapy now also shows promise as a treatment for one type of this syndrome.

After some of his more recent movies, John Travolta has been rumored to be photographed by paparazzi attempting to re-enter the bubble. Good idea.

CHAPTER 8

It's now 4 A.M. and people are drunk, bloated, and exhausted. Leyner is recovering from his Academy Award performance and has his tongue inside the tequila bottle, trying to extract every last drop. He removes his mouth from the bottle and says, "The tongue is God's gift to the human race . . . the

OLD WIVES' TALES

ultimate organ of poetry and pleasuring."

Leyner goes on to say, "The lingua, blessed instrument of storytelling that allows me to continue the tradition of the oral urban legend."

Jeremy, still stinging from his loss in charades, confronts Leyner and says, "I'm so sick of all your stories, my tongue tells me that you should kiss my ass."

Although it's late for most, nothing motivates Leyner more than verbal provocation. He responds with glee. "Ah Jeremy, in medieval times, kissing the ass of a fool's sister was said to cure acne. Have you noticed how clear my skin is lately? Thank your sister for me."

Jeremy leaps at Leyner and the two of them tumble around the floor in a grunting, adolescent flurry of fists and fury. They roll toward the living room and Leyner, although in a seemingly suffocating headlock, is still able to continue his grand historical survey of old wives' tales. "The Visigoths believed that eating juniper berries would make them strong for battle."

Jeremy tries to silence Leyner with a jab to the throat, but in a hoarse voice Leyner adds, "All it did was cause excessive flatulence."

Urban legends and folklore can be the cause of tremendous uncertainty. People often desperately want the record set straight on some of these common myths. So, here you go.

IS IT TRUE THAT YOU HAVE TO WAIT A HALF HOUR AFTER EATING TO GO SWIMMING?

As a child, no time seemed longer than the time spent wait-ing to jump back in the water after a meal. This half hour in hell is not based on science but rather on the minds of nervous parents. There is absolutely no medical evidence that supports waiting thirty minutes before getting back in the pool. Digestion begins immediately when you put food in your mouth, but once the food arrives in your stomach it takes about four hours to process there completely. Food then passes into the small intestine, where it spends an-other two hours, and then on to the large intestine for an-other fourteen. These times vary widely depending on what you eat, so don't set your watch by it.

This doesn't mean that it is safe to eat twelve hamburg-ers and then try to swim the English Channel. Use your head and listen to signals from your body. If you feel pain, cramping, or severe fatigue when swimming, get out, and please don't puke in the pool.

WILL STARING AT AN ECLIPSE MAKE YOU GO BLIND?

Things to avoid staring at:

a woman's cleavage

a large facial mole

a couple making out in public

the sun

The answer to this question is that you probably will not go blind, but staring at an eclipse can indeed cause harm.

The eclipse of the sun on August 11, 1999, put many people at risk of solar retinopathy. Solar retinopathy is the fancy name for damage to the back of the eye caused when radiation from the sun is concentrated by the lens onto the retina. This radiation causes a burn. Solar retinopathy has been studied in medical literature, and surprisingly the damage it causes is not as severe as previously thought. A group of researchers in the United Kingdom studied forty people who experienced eye problems after the August 11, 1999, solar eclipse. It was found that only half suffered from eye discomfort. Only 20 percent of the group of forty reported some damage seven months after the eclipse. These were people who looked directly at the eclipse. It is unclear if these same patients were also staring at cleavage, moles, or amorous couples.

SHOULD YOU STARVE A FEVER AND FEED A COLD?

Or is it feed a fever and starve a cold? Or should you just curl up on the couch, whine like a baby, and call your mommy?

Either way, the answer is no, but there may be some science behind this old wives' tale.

In a study in *Clinical and Diagnostic Laboratory Immunology*, cell biologists in the Netherlands found that starving and feeding affect the immune system in different ways. Scientists looked at healthy volunteers and measured certain chemical messengers in their blood. After a meal, the average level of the chemical that stimulates the body's defense against infections increased by 450 percent. So you should feed a cold and a fever, right?

Not so fast.

Other volunteers, after starving, had high concentrations of another chemical, one that is also associated with the production of antibodies. So the answer is confusing, because it seems as though both starving and feeding a cold or fever can help the immune system.

Like many areas of science, there is no absolutely clear answer here. Our recommendation is that whether you have a cold or a fever, your body needs fluid, rest, and nourishment. If you've lost your appetite, try to drink plenty of fluids and eat whatever healthy food appeals to you and, if in doubt, whine like a baby and call your mommy.

DOES WET OR COLD WEATHER CAUSE A COLD?

A friend once called to ask if she could have caught a cold from touching a goat at a petting zoo. This is not a common question, but many people do ask if any of the following things can cause a cold:

sleeping in front of an open window

getting a chill

sleeping in front of a fan

getting caught in the rain

sleeping with a wet goat in front of a fan in the rain

The answer is no. Cold or wet weather does not cause a cold, but nobody seems to want to accept this.

The common cold is caused by a virus. These viruses are everywhere and it is difficult to avoid them. When you are exposed to someone who has a cold, you are more likely to get ill yourself, so be careful about close contact and definitely wash your hands. Not getting enough sleep or eating poorly can also reduce your resistance to infection. Remember that antibiotics won't fight your everyday cold. Antibiotics work only against bacteria.

To take care of a cold, rest, eat well, and a little chicken soup couldn't hurt. . . .

CAN YOU DIE FROM CHASING POP ROCKS WITH COKE?

From penicillin to Post-its, accidental discoveries have led to many of our most important products. That is how we've come to have Pop Rocks.

Pop Rocks were accidentally invented in 1975 by William Mitchell, a scientist at General Foods. Mitchell was trying to design an instant soft drink when he mixed sugar flavoring and carbon dioxide in his mouth. His startling discovery may or may not have led to the demise of Mikey, the boy from the Life cereal commercials, who as the urban legend goes, liked mixing Coke and Pop Rocks. Supposedly, this deadly mix caused his stomach to rupture.

In 1983, Pop Rocks were taken off the market but recently have come back into fashion like Razzles, Sugar Babies, Charleston Chews, and other retro candy. You should have no problem finding Pop Rocks for your home science experiments. You will surely find that there is no danger in the delicious combination of soda and Pop Rocks.

As for Mikey, he is alive and well and living in post–child star obscurity. As for other child stars, we offer these dramatic gastrointestinally induced death rumors:

1. Gary Coleman (*Diff'rent Strokes*) from snorting Lik-M-Aid and drinking fizzies.
2. Danny Bonaduce (*The Partridge Family*) from ingesting the marshmallows from thirty boxes of Lucky Charms.
3. Erin Moran (*Happy Days*) from choking on Razzles while intoxicated from drinking excessive amounts of Clamato.

CAN LIP BALM BE ADDICTIVE?

When you search the Internet these days, you realize that some people have far too much time on their hands. Conspiracy theorists seem to have it out for many commercial products. Websites claim that lip balm manufacturers lace their products with addictive substances, and lip balm companies use the Internet to refute these claims.

Lip balms may be habit forming, but they certainly aren't addictive. Carmex, which has been accused of containing acid and ground-up fiberglass, contains salicylic acid, which is closely related to aspirin. The salicylic acid works as a pain reliever and as a drying agent but is not addictive. Chap Stick contains white petrolatum (petroleum jelly), lanolin (a wool grease), and padimate O (sunscreen). Nothing addictive there either.

IS IT TRUE THAT LEFT-HANDED PEOPLE ARE SMARTER THAN RIGHT-HANDED PEOPLE?

First we must state that we are right-handed. Second, Billy's wife is left-handed. Mark's wife is ambidexterous and must use both hands to keep him in line.

Medical literature is filled with studies on the effect of right- or left-handedness on oral hygiene, depression, immune function, schizophrenia, enuresis (bed-wetting), longevity, language, asthma, allergies, and injury.

The list continues but none of the evidence is very clear.

In general, the right side of our brain receives input from and controls the left side of our body, and vice versa. Therefore, right-handed people are usually said to be left-brain dominant and left-handers, right-brain dominant. Each brain hemisphere is known to have specialized abilities. The right brain is responsible for visual and spatial skills while the left controls language and speech. Again this does not always hold true. Neuropsychologists have tried to test for intelligence differences between lefties and righties but have found no significant results.

Males are about one and a half to two times more likely to be left-handed than are females. There is no scientific evidence about why this occurs but Mark postulates an evolutionary explanation. Males have developed their left hands to accommodate ambidextrous masturbation.

WILL SLEEPING IN FRONT OF A FAN OR AN OPEN WINDOW CAUSE A STIFF NECK?

Unless you are sleeping beneath an industrial fan that causes your head to wiggle like a bobblehead doll, there should be no problem. This old wives' tale has no scientific basis.

DO MICROWAVES CAUSE CANCER?

This morning I microwaved the milk for my coffee, and a few hours later I heated up some lasagna for lunch. If what you read on the Internet is true, I should have about twelve more hours to live.

But no studies have proven modern microwave usage to be harmful. Much of the fear about the cancer-causing agents of microwaves has to do with radiation. Basically anything that moves is radiation, including visible light, ultraviolet rays, X rays, and microwaves. Ionizing radiation, such as X rays, have enough localized energy to do chemical damage to the molecules they hit. Nonionizing radiation, such as microwaves, do not damage molecules.

One possible danger with microwaves is that heated products can explode even after they are removed from the microwave. Exploding eggs are specifically dangerous. Many injuries have been reported and some doctors in the United Kingdom have even pressed for warning labels.

WILL USING A CELL PHONE GIVE YOU A BRAIN TUMOR?

Wireless phones (including cell phones) use radiofrequency energy, also known as radio waves. It is not believed that wireless phones are harmful, but the research in this arena has only been conducted recently, so the real negative effects of cell phone usage remain unknown for now.

WILL A PLATE IN YOUR HEAD SET OFF A METAL DETECTOR IN THE AIRPORT?

This question makes me think of the scene from *High Anxiety* when Mel Brooks, playing Dr. Richard Thorndyke, passes through airport security with a gun. As the metal detector beeps, he bursts out, "Is this a game show? What did I win, a Pinto? I beeped! Take me away! Take me back to Russia! I beeped! The mad beeper is loose!"

If you have a titanium plate in your head, a pacemaker, plates and screws for a broken bone, or an artificial implant, this too could happen to you. The size of the implant and the sensitivity of the device will determine whether you are turned into the mad beeper. Don't worry, the metal detector won't hurt you.

IS IT DANGEROUS TO HOLD IN A SNEEZE?

The old wives' tale warns us that if you hold in a sneeze, your head might explode. That won't happen, but you can do yourself some harm.

A sneeze is a very complicated thing that involves many areas of the brain. A sneeze is a reflex triggered by sensory stimulation of the membranes in the nose, resulting in a co-ordinated and forceful expulsion of air through the mouth and nose. The Guinness Book of World Records reports the longest sneezing bout ever recorded was that of a school-girl from the United Kingdom. She started sneezing on January 13, 1981, and didn't stop sneezing for 978 days.

The air expelled by sneezes is said to travel up to one hundred miles per hour, and an unimpeded sneeze sends two to five thousand bacteria-filled droplets into the air. Holding in a sneeze potentially can cause fractures in the nasal cartilage, nosebleeds, burst eardrums, hearing loss, vertigo, detached retinas, or temporary swelling called facial emphysema. Therefore it is best to let your sneeze fly, but please cover your nose and mouth.

CAN YOU SWALLOW YOUR TONGUE?

Several years ago at the Columbus Circle entrance to Central Park, I came upon someone having a seizure in the street. As I attempted to help the patient, someone from the crowd reached into the nearby garbage can and insisted that I stick the dirty spoon he had found into the person's mouth to keep him from swallowing his tongue. The guy with the spoon didn't seem to be impressed with my medical degree and "politely" told me that I didn't know what the (insert vulgar NY expression here:___) I was talking about.

This is not an uncommon belief, but it is not possible to swallow your tongue. The tongue *can* block the opening of the airway and one of the first things that you are taught in basic life support is that if someone is having difficulty breathing, you should tilt his or her head and lift the chin. This helps to remove the tongue as an obstruction. If you do come upon someone who is having a seizure, just make sure that he is safe and won't hurt himself. Do not put a bacteria-covered spoon in the mouth. Call for help and before you know it, the seizure will probably stop on its own.

CHAPTER 9

I can't believe it's not over yet. I feel as though this evening has taken years off my life. Leyner and Jeremy have been separated, and there are only a few stragglers left picking at the remnants of Eloise's glorious buffet.

Even Leyner seems beaten down from a combi-

nation of toxic tequila, amorous adventures, and verbal violence. He is leaning on the credenza and says to me as he agonizingly stretches his neck, "I used to be able to drink, womanize, and brawl and come out of it all as fresh as a daisy. Now I feel limp and shriveled like a rotting clump of stinkweed."

Leyner stands and arches his back uncomfortably. "Did I mention my prostate feels a little swollen?"

With that, I turn and exit the party.

There are many advantages to getting older—early bird specials, senior citizen discounts, the fact that people don't ask you to help move a sofa up a flight of stairs, and getting away with saying whatever the hell pops into your head. But there are some perplexing changes ahead for all of us. . . .

5:33 P.M.

Gberg: Maestro.

Leyner: Hey you . . . give me five minutes
(at most) . . . go get something . . . then
we'll work.

Gberg: Surely.

5:45 P.M.

Leyner: You there?

Gberg: Yes, sir.

Leyner: What should we do?

5:50 P.M.

Gberg: Light this piece of shit on fire
and go drink ourselves silly.

Leyner: Brilliant idea.

Gberg: Or we can talk about the health
alert that I just received about mycobacterium bovis in U.S.-born children.

Leyner: What the hell is that?

Gberg: You are so filled with all of this
medical knowledge, I was hoping you could
fill me in.

Gberg: I am waiting.

Leyner: What is it? Some mushroom thing,
some fungal thing?

Leyner: Some fungal cow thing.

Gberg: You are getting warm with the cow.

Gberg: Sounds perverse.

Gberg: Give up?

Leyner: I give up.

Gberg: Similar to TB (mycobacterium tuber-culosis).

5:55 P.M.

Leyner: Duh . . . should have known that.

Gberg: Infection that you can get from the milk of infected cattle.

Gberg: Unpasteurized cheese and stuff.

Leyner: People should just drink cow's blood . . . like the Masai . . . that would solve the problem of mycobacterium and lactose intolerance.

Gberg: You should run the Department of Health.

Leyner: Thank you.

Gberg: And Homeland Security.

Leyner: Homeland Senility.

Gberg: We have to finish this book first.

....

IS IT TRUE THAT YOU LOSE TASTEBUDS AS YOU GET OLDER?

You finally have time and money to relax, travel, and have a good meal. The bad news is you probably can barely taste this meal.

Starting at age forty-five, tastebuds begin to lose much of their sensitivity. Older people often lose their ability to sense bitter or salty flavors altogether. You start your life with about nine thousand tastebuds and in old age you have less than half of that.

You want some more bad news? Aging also causes decreased hearing, sight, smell, and touch.

WHY DOES HAIR TURN GRAY?

All the hairs on our head contain pigment cells that contain melanin. Pigment cells in our hair follicles gradually die as we age. The decrease in melanin causes the hair to become a more transparent color like gray, silver, or white.

Premature gray hair is hereditary, but it has also been associated with smoking and vitamin deficiencies. Early onset of gray hair (from birth to puberty) can be associated with medical syndromes including dyslexia.

A more interesting question is why old ladies insist on trying to cover up their gray hair with bright blue hair dye.

WHY DO YOU SHRINK AS YOU GET OLDER?

Some of us don't have a lot of inches to lose. In height, that is.

Unfortunately, we all will get a little shorter as we age. This takes place over many years, and ultimately we all lose an inch or so. Gravity is responsible for some of this height loss. You lose muscle and fat as you age and gravity weighs down especially on the bones in your spine and may cause compression. This explains why all those senior drivers in Florida can barely see over the dashboard.

WHY DO OLD LADIES GROW BEARDS?

The easy answer would be to work in a carnival, but it is actually not that simple.

At menopause, the ratio of male hormones, or androgens, to estrogen begins to change. This can produce mild increases in facial hair. The amount or thickness of facial hair is hereditary and how thickly hair follicles are distributed throughout the skin is determined at birth. Some ethnic groups or nationalities are more likely to develop facial hair than others.

Some medical conditions can cause excessive hair growth, so it is always wise to check with your doctor especially if you are a woman experiencing five o'clock shadow.

DO YOUR EARS CONTINUE TO GROW AFTER THE REST OF YOUR BODY STOPS GROWING?

Prince Charles may worry about this very question.

There are definitely some changes in the face that occur with aging. First some facial muscle tone is lost, causing that saggy look. Then you get the dreaded double chin. The nose can also lengthen a bit, and the skin on the face becomes thin, dry, and wrinkled. Then there are longer, thicker eyebrows and gray hair. We haven't even mentioned droopy eyes, receding gums, missing teeth, and last but not least— bigger ears. Yes, your ears do continue to grow as you age, but only slightly. This is probably due to cartilage growth.

What a list of wonderful things to look forward to as we enter our golden years.

WHY DO YOU NEED LESS SLEEP WHEN YOU GET OLDER?

Actually, you do not need less sleep as you get older.

The body's sleep requirements remain constant throughout our lives. The average total sleep time, however, actually increases slightly after age sixty-five. This sounds like something to be excited about, but not really. The problem is that as you age, you have more difficulty falling asleep. Sleep for the elderly is also interrupted by such factors as leg cramps, sleep apnea, and medical or psychiatric illness.

Normal sleep consists of two major states: REM (rapid eye movement) sleep and NREM (non-REM) sleep. NREM sleep is divided further into four sleep stages. A healthy night's rest is generally comprised of 20 percent REM and 80 percent NREM. As you age, this distribution is changed.

WHAT'S UP WITH THE EAR HAIR?

You lose the hair where you want it, and gain it in all those other unsightly places. Bushy eyebrows, excessive nasal hair, and hairy ears certainly don't make you anxious to get older, do they?

Sometimes the excessive growth of hair on the ears is genetic and is linked to the Y chromosome, the sex chromosome found only in males, which explains why you don't see many hairy-eared females, except in *The Lord of the Rings* movies.

And what would this excess hair growth be without a competition? The *Guinness Book of World Records* record for the longest ear hair was broken again in 2002. A seventy-year-old from Tamil Nadu state in India, Anthony Victor, broke the record with his ear hair measuring 11.5 centimeters.

DO YOUR NAILS OR HAIR GROW AFTER YOU DIE?

Human nails and hair do not grow after death. The fact of the matter is that after you die, your body starts to dry out, creating the illusion that your hair and nails are still growing as the rest of you shrivels up.

WHAT ARE AGE SPOTS?

Age spots are also known as sunspots or lentigines. They are flat, brown discolorations of the skin that usually occur on the back of the hands, neck, and face of people older than forty years of age.

Age spots are caused by an increased number of pigment-producing cells in the skin. As our skin becomes thinner with age, it also becomes more translucent, which makes these spots more obvious. Age spots are caused by the skin being exposed to the sun over many years and are a sign of sun damage. They are not harmful and do not represent skin cancer.

IS LIFE SPAN DETERMINED STRICTLY BY GENETICS?

How we age as individuals is a complex interaction of genetic and environmental factors. There have been many studies that try to assess how much of our longevity is determined by our genes. Scientists have known for several years that people who live the longest often have children who also have long life spans. The life spans of adoptees seem to be more closely correlated to those of their birth parents than to those of their adoptive parents. One study of twins reared apart suggests about a 30 percent role for heredity in life span, but others say the influence is even smaller. Recent research seems to indicate that the process of aging and life span may be determined by your mother's X chromosome. Incessant maternal nagging, however, could reverse any beneficial genetics.

....

Leyner: I spoke to my grandmother last night . . . but she thought I was her son, my uncle.

Leyner: So I had the conversation with her as my uncle.

Gberg: Deceptive but useful.

Leyner: It was simpler than disabusing her of the mistaken identity.

Gberg: How old is she now?

Leyner: 96.

Leyner: I'm going to be 49 on Tuesday.

Gberg: I can't imagine you as a 96-year-old.

Leyner: Feels like I'm teetering on the precipice of something major.

11:40 A.M.

Gberg: What, the publication of this masterpiece?

Gberg: Do you have plans for Tuesday?

Leyner: Yes! We should start writing Nobel acceptance speeches now. I just booked a ticket for Stockholm on Travelocity. This book will do for us what "The Little Red Book" did for Mao.

Leyner: Speaking of not going gently into that good night . . . Mao is my model for groovy aging.

Gberg: That's the first time I've heard Mao and groovy in the same sentence.

Leyner: Smoked five packs a day . . . never brushed his teeth (just rinsed with green tea) . . . AND cavorted every night with three or four young Red Guard hoochie girls.

Gberg: I never knew he was such a player. He probably would have loved to party with Kim Jong-il.

11:45 A.M.

Leyner: Mao got more action in a week than Kim Jong-il will have in a lifetime.

Gberg: Those were the days, my friend.

Leyner: Our next book should be "The Sex Lives of Asian Despots."

Gberg: We need a catchy title, like . . .

Gberg: . . . "Despots and Sexpots."

Leyner: That's pretty good—maybe we should just call this book "Despots and Sexpots."

••

CAN TAKING VITAMIN C HELP YOU LIVE LONGER?

Dr. Linus Pauling, a two-time Nobel Prize winner, took high doses of vitamin C for almost forty years and died at the ripe old age of ninety-three. He believed that his life was prolonged for twenty years because of his high vitamin C intake. Sounds great, and Pauling certainly makes a valid argument but unfortunately there isn't any strong evidence to support this claim.

Vitamin C and vitamin E, often referred to as antioxidant vitamins, have been suggested to prevent cell damage in humans, thereby lowering the risk of certain chronic diseases including high blood pressure, stroke, and asthma. Many studies have tried to prove this, but with no obvious results.

Very few side effects exist with vitamin C, so there is no major downside to adding a vitamin C supplement to your daily regimen, though occasional reports of nausea, heartburn, gas, or diarrhea are noted with higher doses. A more prudent approach is to eat a balanced diet including fruits and vegetables and avoid dangerous activities like smoking.

IS THERE SUCH A THING AS MALE MENOPAUSE?

The concept of male menopause and the need for hormone replacement therapy as a treatment is a highly controversial topic. The many names that are used to describe these changes in an aging male include andropause, viropause, male climacteric, ADAM (Androgen Decline in the

Aging Male) syndrome, Aging Male Syndrome (AMS), or late-onset hypogonadism. Some wives prefer to call it a midlife crisis. Even though we don't like to admit it, they may be right.

Male menopause is thought to be due to a decrease in testosterone. In contrast to the female menopause, the process in men is characterized by slow onset and slow progression. There is a progressive reduction in testicular function in men between the ages of twenty-five and seventy-five. During this time, the amount of available testosterone can fall by almost 50 percent. This is not a fixed number and there is a great deal of variation among individual men.

The symptoms of this syndrome are lethargy, or fatigue, depression, increased irritability, mood swings, decrease in lean muscle, increase in fat, decreased libido, and difficulty in attaining and sustaining erections. Many men now receive treatment with testosterone, and report improved symptoms. There is a downside to this treatment as it can increase the risk of prostate cancer and atherosclerosis.

Many experts believe that this syndrome is more likely to be the result of nonhormonal explanations or the normal aging process. Lifestyle factors such as alcohol and drug use, medications, marital problems, financial problems, and stress in general all may have a role. Doesn't sound much different than a midlife crisis.

11:15 A.M.

Leyner: I'm going to give my daughter some Benadryl and I'll be right there.

Leyner: Be right back.

Gberg: Okay—

Gberg: So, you medicate your daughter in order to get work done.

11:20 A.M.

Leyner: How dare you suggest such a thing— pharmaco-parenting!

Gberg: A little Benadryl beats a little Ritalin.

Gberg: What, are you too offended to respond?

Leyner: I'm stretching to gear up for this session. . . .

Gberg: I am having a cup of coffee and trying to jump-start my lethargic body.

Gberg: I can't believe I have to work 4-12.

Leyner: How are you feeling? Any better?

11:25 A.M.

Gberg: Tired, didn't sleep all that much.

Gberg: Just getting old, I guess.

Gberg: Male menopause.

Leyner: Shouldn't old people sleep better, sort of like dry runs for death?

Gberg: Manopause.

Leyner: I think I'm getting younger from the neck down and rapidly achieving senescence from the neck up.

Gberg: Please just stay above the waist—it's too early to hear anything more.

Leyner: How come lawyers never use male menopause as a mitigating defense?

Gberg: What, a hot flash causing road rage?

11:30 A.M.

Leyner: A hot flash causing an otherwise rational person to decide that murder makes more practical sense than divorce . . .

Leyner: Aren't we entering into a world in which old-age homes will be populated with withered hags with enormous perfect boobs and hunched drooling men struggling to haul their pec implants around the joint?

Gberg: Both covered in faded tattoos that now resemble ancient cave paintings.

Leyner: Exactly!!!!!

Gberg: But at least the blue hair will seem somewhat funky rather than freaky . . .

Gberg: Or maybe not.

WHY ARE OLDER PEOPLE SUCH BAD DRIVERS?

Our population is aging. Persons sixty-five and older now represent about 12 percent of the total U.S. population. In 2050 there should be about 86.7 million seniors in the United States, comprising 21 percent of the total population. I hope that they aren't driving in front of me. Wait a minute . . . that is going to be *me*!

It is true that older drivers are more likely to get in multiple vehicle accidents, more likely to get traffic citations, and more likely to get seriously injured in a crash than younger drivers. One theory is that they tend to overestimate their abilities and are less able to compensate for their mistakes.

There are also some medical reasons that older drivers are less adept behind the wheel. First, seniors need to contend with common vision problems. Loss of visual acuity (especially night vision) is a problem and even when corrected there is usually still a loss of peripheral vision, contrast vision, and depth perception. Hearing loss makes drivers less likely to hear important cues such as sirens, horns, or screeching tires, too. Restricted mobility, weakness, and decreased reaction time also contribute to driving problems.

IS THERE REALLY A WRINKLE CURE?

The only way to prevent wrinkles is to avoid aging or to freeze yourself like Austin Powers. Otherwise, you will have to stick with sunscreen and moisturizers. To reduce or improve wrinkles there are several options.

Tretinoin (Retin-A) is the only topical medication that has been clearly proven to improve wrinkles in controlled clinical studies. It is also the only FDA-approved medical treatment for this purpose. Tretinoin increases sensitivity to sunlight, therefore sun avoidance, protective clothing, and sunscreen are recommended when used. Side effects include peeling, dry skin, burning, itching, and redness.

Hydroxy acids are also present in many over-the-counter creams and there is some evidence that they may help with minor wrinkles. There is great variability in the amount and type of hydroxy acid in different products and therefore variable efficacy.

Other more drastic treatments include chemical peels and laser resurfacing. The short answer, though, is not really.

CAN ALUMINUM CAUSE ALZHEIMER'S?

If aluminum caused Alzheimer's, wouldn't the Tin Man have needed a brain rather than a heart?

Aluminum intake is unavoidable. It comes primarily from food, drinking water, and pharmaceuticals, like antacids. It occurs both naturally and as an additive. It also can leach into food from the pans we use. Aluminum has been linked with Alzheimer's disease since the 1960s. For almost every study in favor of a connection, there has been a study against it. Like many scientific theories, there remain many unanswered questions. The majority of scientists now believe that if aluminum plays any role in Alzheimer's at all, it is very small.

What does this mean for us? It means we can relax. Aluminum is the third most common element in our world after oxygen and silicon, so it would be extremely difficult to entirely avoid aluminum. If you do choose to try and avoid aluminum, you can drink filtered water, avoid aluminum-containing antiperspirants, and be careful when cooking acidic or basic foods in aluminum-containing cookware.

If you are going to be an anti-aluminum crusader, we hope that you will be consistent. There is nothing more annoying than people preaching to you about eating organic while they are smoking cigarettes.

12:55 P.M.

Gberg: Are you there?

Gberg: Shall we call it quits?

Leyner: When I get sufficiently old and incontinent, I want to be taken to the doctor in one of those pet-carrying cases . . . like a cat . . . and left there.

Leyner: Then I want to be boiled down and canned.

Leyner: My version of a pharonic memorial . . . instead of mummification . . . canning . . .

Gberg: And then your entire family can keep a can of you on a back shelf next to the Spam.

Leyner: Exactly . . . a fitting tribute.

Leyner: That seems to wrap all this up: aging, family, cannibalism, and the pursuit of health . . .

Gberg: We have come full circle.

ABOUT THE AUTHORS

MARK LEYNER is the author of *My Cousin, My Gastroenterologist; Tooth Imprints on a Corn Dog; I Smell Esther Williams; Et Tu Babe;* and *The Tetherballs of Bougainville.* He has written scripts for a variety of films and television shows. His writing appears regularly in *The New Yorker, Time,* and *GQ.*

BILLY GOLDBERG, M.D., is an emergency medicine physician on faculty at a New York City teaching hospital. He is also a writer and artist whose paintings have been exhibited in New York City.